METHODS IN MOLECULAR BIOLOGY™

Series Editor
John M. Walker
School of Life Sciences
University of Hertfordshire
Hatfield, Hertfordshire, AL10 9AB, UK

For other titles published in this series, go to
www.springer.com/series/7651

Protocols for Adult Stem Cells

Edited by

Irina M. Conboy

Department of Bioengineering, University of California, Berkeley, CA, USA

David V. Schaffer

Department of Chemical Engineering, Bioengineering, and Neuroscience, University of California, Berkeley, CA, USA

Mary Helen Barcellos-Hoff

Department of Cancer Biology, Life Sciences Division, Lawrence Berkeley National Laboratory, Berkeley, CA, USA

Song Li

Department of Bioengineering, University of California, Berkeley, CA, USA

Humana Press

Editors
Irina M. Conboy
Department of Bioengineering
University of California
Berkeley, CA
USA
iconboy@berkeley.edu

Mary Helen Barcellos-Hoff
Department of Cancer Biology
Life Sciences Division
Lawrence Berkeley National Laboratory
Berkeley, CA
USA
mhbarcellos-hoff@nyumc.org

David V. Schaffer
Department of Chemical Engineering,
Bioengineering, and Neuroscience
University of California
Berkeley, CA
USA
schaffer@berkeley.edu

Song Li
Department of Bioengineering
University of California
Berkeley, CA
USA
song_li@berkeley.edu

ISSN 1064-3745 e-ISSN 1940-6029
ISBN 978-1-4939-6104-7 ISBN 978-1-60761-063-2 (eBook)
DOI 10.1007/978-1-60761-063-2
Springer New York Dordrecht Heidelberg London

Printed on acid-free paper

Humana Press is a part of Springer Science+Business Media (www.springer.com)

Preface

Although embryonic stem cells currently enjoy the public limelight and show great promise for cell based medical therapies, it is the adult stem cells which are responsible for the body's natural ability to fight disease, heal and recover, or fail and succumb to various maladies. The study of mammalian adult stem cells has surged recently, most likely from a maturation of stem cell studies in the classical developmental model organisms and in hematopoeisis. All the tissues of the body examined so far are generated and regenerated from stem cells, it has been an important first step to adapt or devise new methods to identify and obtain these cells in quantity and purity for further study. Culture techniques have been optimized for managing the growth and differentiation of stem cells in vitro; as some stem cells are pluripotent, often the method is to guide the fate of such cells among the possible differentiation fates. Much of this work, and that in the classical model organisms, has helped define the aspects of the stem cell environment or niche that are crucial for both growth and differentiation, and these studies have moved in vivo at increasingly higher resolution. Importantly, the in vivo niche is a current target for bioengineering the matrix and signaling factors.

Herein, we present methods for studying six types of mammalian stem cells, mammary, neural, mesenchymal, endothelial, dendritic, and muscle. We hope the approaches described in these chapters will assist you in obtaining, characterizing, and studying these cells or for adaptation to your favorite stem cells. One final note, should you find the information herein valuable and it helps you to make great discoveries in the understanding of stem cell biology or application, by all means please let us, the authors, know! The best way is of course to publish your findings, but falling short of that, please feel free to ask questions or clarifications. Especially for this reason, our contact information is provided at the end of this book.

Berkeley, CA

Dr. Irina M. Conboy
Dr. David V. Schaffer
Dr. Mary Helen Barcellos-Hoff
Dr. Song Li

Contents

Contributors

ANSHU AGRAWAL • *Division of Basic and Clinical Immunology, Department of Medicine, University of California, Irvine, CA, USA*
SMITA AGRAWAL • *Department of Chemical Engineering, University of California, Berkeley, CA, USA*
MARY HELEN BARCELLOS-HOFF • *Langone School of Medicine, New York University, New York, NY 10016, USA*
MASSIMILIANO CERLETTI • *Section on Developmental and Stem Cell Biology and Department of Pathology, Joslin Diabetes Center and Harvard Stem Cell Institute, Boston, MA, USA*
TATHAGATA CHAUDHURI • *Pennsylvania Muscle Institute, University of Pennsylvania, Philadelphia, PA, USA*
IRINA M. CONBOY • *Department of BioEngineering, University of California, Berkeley, CA, USA*
MICHAEL J. CONBOY • *Department of BioEngineering, University of California, Berkeley, CA, USA*
DENNIS E. DISCHER • *Graduate Groups in Cell and Molecular Biology, Physics, and Chemical & Biomolecular Eng., Pennsylvania Muscle Institute, University of Pennsylvania, Philadelphia, PA, USA*
RODRIGO FERNANDEZ-GONZALEZ • *Development Biology Program, Sloan-Kettering, Institute, Memorial Sloan-Kettering Cancer Center, New York, NY 10065, USA*
CRAIG HASHI • *Department of Bioengineering, University of California, Berkeley, CA, USA; Joint Graduate Group in Bioengineering, University of California, San Francisco and Berkeley, Berkeley, CA, USA*
IRINEU ILLA-BOCHACA • *Langone School of Medicine, New York University, New York, NY 10016, USA*
STEVE KHACHI • *Department of Anesthesia and Perioperative Care, Center for Cerebrovascular Research, University of California, San Francisco, CA, USA*
SONG LI • *Department of Bioengineering, University of California, Berkeley, CA, USA*
BRANDI K. ORMEROD • *J. Crayton Pruitt Family Department of Biomedical Engineering, University of Florida, Gainesville, FL, USA*
CARLOS ORTIZ-DESOLORZANO • *Morphology and Imaging Group and Cancer Imaging Laboratory, Center for Applied Medical Research, University of Navarre, Pamplona, 31008 Navarre, Spain*
JENNIFER S. PARK • *Department of Bioengineering, University of California, Berkeley, CA, USA; Joint Graduate Group in Bioengineering, University of California, San Francisco and Berkeley, CA, USA*
JOSEPH PELTIER • *Department of Chemical Engineering, University of California, Berkeley, CA, USA*

FLORIAN REHFELDT • *Department of Physics, University of Pennsylvania, Philadelphia, PA, USA*
MATTHEW J. ROBERTSON • *Department of Chemical Engineering, University of California, Berkeley, CA, USA*
DAVID V. SCHAFFER • *Department of Chemical Engineering, University of California, Berkeley, CA, USA*
DAWNE N. SHELTON • *Department of Surgery, University of Utah, Huntsman Cancer Institute, Salt Lake City, UT 84112, USA*
H. LEE SWEENEY • *Department of Physiology, Pennsylvania Muscle Institute, University of Pennsylvania, Philadelphia, PA, USA*
AMY J. WAGERS • *Section on Developmental and Stem Cell Biology and Department of Pathology, Joslin Diabetes Center and Harvard Stem Cell Institute, Boston, MA, USA*
BRYAN E. WELM • *Department of Surgery, University of Utah, Huntsman Cancer Institute, Salt Lake City, UT 84112, USA*
GUO-YUAN YANG • *Department of Anesthesia and Perioperative Care and Department of Neurological Surgery, Center for Cerebrovascular Research, University of California, San Francisco, CA, USA*
YIQIAN ZHU • *Department of Anesthesia and Perioperative Care, Center for Cerebrovascular Research, University of California, San Francisco, CA, USA*

Chapter 1

In Situ Analysis of Cell Populations: Long-Term Label-Retaining Cells

Rodrigo Fernandez-Gonzalez, Irineu Illa-Bochaca, Dawne N. Shelton, Bryan E. Welm, Mary Helen Barcellos-Hoff, and Carlos Ortiz-de-Solorzano

Abstract

The mammary gland consists of an epithelial ductal tree embedded in a fat pad. Adult mammary epithelium has been demonstrated to have outstanding regenerative potential, consistent with the presence of resident, adult stem cells. However, there are currently no *bona fide* markers to identify these cells within their tissue context. Here, we introduce long-term label retention as a method to investigate the location of quiescent cells (a property attributed to adult stem cells) in situ. Long-term label retaining cells divide actively during tissue development and remain quiescent at homeostasis. These two properties have been attributed to adult stem cells. Therefore, label-retaining cells can be used to identify populations that contain stem cells. We describe the materials and methods necessary to identify and image mammary label-retaining cells, to carry out morphometric analysis on these cells and to map their distribution of the mammary epithelium. The morphometric and spatial analyses described here are generally applicable to any mammary cell populations, and will therefore be useful to characterize mammary stem cells once *bona fide* mammary stem cell markers become available.

Key words: Bromodeoxyuridine labeling, Osmotic pump, 3D reconstruction, Nuclear morphology, Spatial analysis

1. Introduction

In neonate mammals, the mammary epithelium consists of a primordial sprout branching from the specialized epidermis that forms the nipple (1). The mature ductal tree develops during puberty under the influence of the steroid hormones estrogen and progesterone acting through their cognate nuclear receptors (estrogen receptor and progesterone receptor). Driven by the hormonal stimulus, the mammary ducts invade and fill the fat pad

Irina M. Conboy et al. (eds.), *Protocols for Adult Stem Cells*, Methods in Molecular Biology, vol. 621,
DOI 10.1007/978-1-60761-063-2_1, © Springer Science+Business Media, LLC 2010

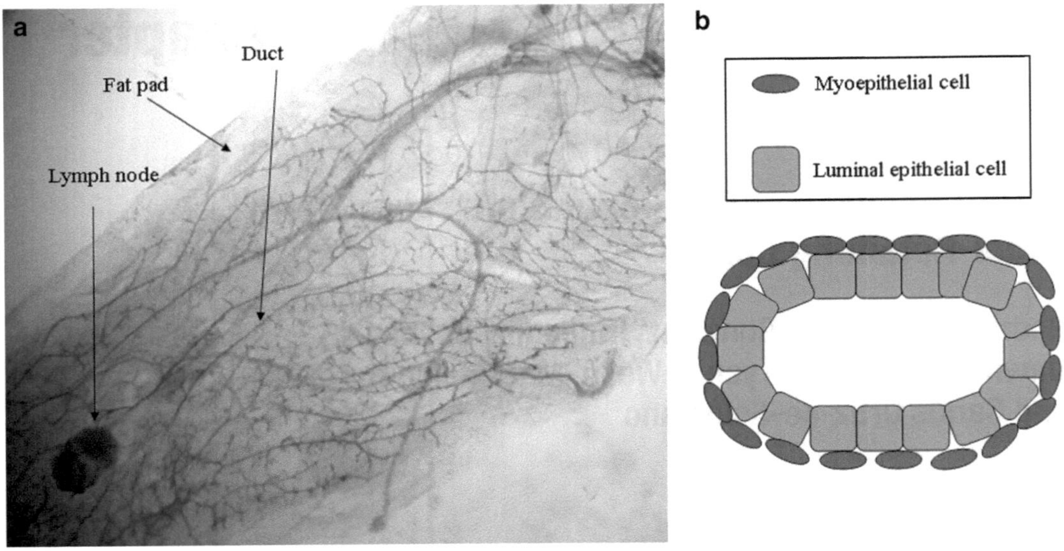

Fig. 1. Mammary gland architecture. (**a**) Whole mount preparation of a mouse mammary gland. The tissue consists of a ductal tree with three levels of branching embedded in a fat pad. (**b**) Cartoon representing a section through a mammary duct. The duct is lined by luminal epithelial cells, which are in turn surrounded by a mesh of contractile, myoepithelial cells

in which they are embedded (2, 3). During pregnancy, the mammary gland develops secretory, alveolar structures from the ductal tree that drains into the nipple (Fig. 1a). This secretory capacity is ablated during the involution that ensues upon weaning but is regenerated with each subsequent pregnancy. The mammary epithelium consists of a monolayer of luminal epithelial cells (Fig. 1b) surrounded by differentiated, contractile myoepithelial cells (4).

The postnatal development of the gland and the massive side branching and alveolar growth that the mammary gland undergoes upon each pregnancy confirm the existence of a *reservoir* of proliferative potential in the mammary epithelium. Moreover, this growth potential is preserved through multiple pregnancies through the fertile life of the animal. The replicative and differentiating potential of the adult mammary gland is consistent with the presence of a population of tissue-specific stem cells.

ASC have been shown to form clusters in several epithelial tissues, like the skin (5–7) or the small intestine (8, 9). ASC behavior is thought to be highly dependent on the signals coming from their local cellular environment, commonly referred to as the niche (10). Thus, understanding the role of ASC in a given organ depends on being able to properly define the location of the ASC in the organ as well as the local architecture of the niche. However, markers that are reliably enriched for ASC content from cell suspensions, such as integrins (11, 12), are frequently not useful in situ because of their broad histological distribution. Additionally, the conditions that the cells withstand in vitro can affect their expression of certain proteins (13). Thus, disruption of the microenvironment renders characterization of the spatial distribution of ASC within the tissue impossible.

Further motivation for defining the distribution of mammary progenitors stems from the potential role of ASC in neoplasia. Several transgenic mouse mammary models that are more susceptible to carcinogens or spontaneous transformation have increased ASC numbers based on outgrowth formation frequency, but also exhibit disrupted morphogenesis (14, 15). One might then ask whether the intrinsic cell fate decisions are perturbed, or whether signaling from the niche is disrupted. And this can only be assayed by investigating ASC within their tissue context.

The study of long-term, label-retaining cells (LRC) provides a means to investigate ASC populations in situ. LRC are recognized by their proliferative activity during tissue development or regeneration, when a thymidine analog, the label, is administered to mark dividing cells; and by their relative quiescence at homeostasis, which preserves the label within the nuclei of LRC for a long period of time or chase, while proliferating cells lose it. Both of these properties have been attributed to ASC. In fact, LRC have been successfully used as ASC markers in intestinal crypts (8), hair follicles (7), epidermis (16) and prostate (17). Functional assays or bona fide ASC markers have confirmed that LRC contain a subpopulation of ASC (9, 18–20). In the intestine, however, recent evidence suggests that LRC do not correspond to the stem cell population, which is actively proliferating (21). However, it is not yet clear whether this proliferating population corresponds to ASC or to an unrestricted progenitor responsible for tissue regeneration (22). In the mammary gland, some LRC have ASC properties (15, 23, 24). Therefore, in the absence of definitive markers to recognize mammary ASC in situ, LRC can be used as a surrogate marker to identify populations that contain some cells with stem/progenitor cell properties (25).

We have further developed LRC analysis by integrating frequency data with additional information about location, morphological features, and neighbor relationships at multiple organizational scales. By using semiautomated image analysis, one can calculate a large number of nuclear properties exhibited in situ and conduct multiscale in situ cell sorting by defining distinct features of LRC compared to non-LRC and/or to cells that express specific proteins, like hormone receptors. We also developed an automated algorithm to quantitatively determine the degree of spatial clustering of a cell population and its statistical significance (26). Using this value, it is possible to compare different morphological structures to determine whether LRC clustering preferentially occurs in any one of them, as observed in the hair follicle bulge (7) or the bottom of the crypts of the small intestine (27).

Here we present protocols that combine classic immunofluorescence and wide-field microscopy with novel quantitative, image-based measurements at multiple scales (single cell, population of ducts, entire tissue) to characterize the morphology and spatial patterns of distribution of mammary LRC. We show how to quantify morphological features and expression of different markers in single cells to investigate the heterogeneity of LRC. We finally demonstrate how to measure the distribution of LRC along the ductal tree.

2. Materials

2.1. Bromodeoxyuridine and Pump Preparation

1. Bromodeoxyuridine (BrdU, Sigma #B5002, 1 g).
2. Sterile phosphate-buffered saline (PBS, EMD #6505).
3. Pumps (Alzet osmotic pumps, #1002, 100 μl/pump). Twice as many as mice (see Note 1).
4. Glass vial and 1.5 ml eppendorf tubes.
5. Magnet and stirring plate.
6. Tube warmer.
7. 1 ml syringes and 27″-gauge needles.
8. Cotton swabs and paper towels.
9. Balance.

2.2. Pump Implants

1. Mice. Order 20-day-old females of at least 10 g of weight and no more than 12 g a week ahead and have them delivered the day before the procedure, so that they are 21 days old on the day of the procedure (see Note 2).
2. BrdU-filled pumps. Prepared as described in Subheading 3.1.
   a. Stock solution: dissolve 25 g of 2,2,2-tribromoethanol (Sigma T48402, 100 g) into 15.5 ml tert-amyl alcohol (Fisher A730-1). For this you will have to stir for 12 h in a foil-wrapped bottle (avertin is light-sensitive). Wash the magnet that you will use with hot water before starting. Store at room temperature.
   b. 0.9% saline: 0.9 g of NaCl in 100 ml deionized or distilled water. Filter sterilize with a 0.2 μm (or 0.22 μm) filter.
   c. Working solution: stir 0.5 ml stock solution into 39.5 ml of 0.9% saline for 12 h. Wash the magnet that you will use for this with hot water before starting to stir. Store at 4°C (see Note 3).

   d. Dosage: avertin, 0.2 ml/10 g body weight; ketamine, 0.02 ml/10 g body weight.
3. Avertin (ketamine can also be used as anesthetic).
4. Tags and tagging machine.
5. Wound clips, clipping machine, and clip remover.
6. Surgery instrumentation: scissors and two forceps.
7. Corkboard (with rubber bands, no pins!).
8. Cotton swabs and paper towels.
9. Clean extra cages.
10. Lamp.

2.3. Pump Explants

1. Mice with pumps implanted as described in Subheading 3.2.
2. Avertin (ketamine can also be used as anesthetic).
   a. Stock solution: dissolve 25 g of 2,2,2-tribromoethanol (Sigma T48402, 100 g) into 15.5 ml tert-amyl alcohol (Fisher A730-1). For this you will have to stir for 12 h in a foil-wrapped bottle (avertin is light-sensitive). Wash the magnet that you will use with hot water before starting. Store at room temperature.
   b. 0.9% saline: 0.9 g of NaCl in 100 ml deionized or distilled water. Filter sterilize with a 0.2 μm (or 0.22 μm) filter.
   c. Working solution: stir 0.5 ml stock solution into 39.5 ml of 0.9% saline for 12 h. Wash the magnet that you will use for this with hot water before starting to stir. Store at 4°C (see Note 3).
   d. Dosage: avertin, 0.2 ml/10 g body weight; ketamine, 0.02 ml/10 g body weight.
3. Wound clips, clipping machine, and clip remover.
4. Surgery instrumentation: scissors and two forceps.
5. Corkboard (with rubber bands, no pins!).
6. Cotton swabs and paper towels.
7. Clean extra cages.
8. Lamp.

2.4. Tissue Collection

2.4.1. Tissue Collection

1. CO_2 source.
2. Mice treated as described in Subheadings 3.2 and 3.3.
3. Corkboard and rubber bands.
4. 70% ethanol.
5. Paper towels.
6. Surgery instrumentation: scissors and two forceps.

2.4.2. Formalin Fixation

1. Tissue cassettes.
2. Waxed or glassine paper (regular white paper can also be used).
3. 10% buffered formalin (Fisher #UN3334).
4. 70% ethanol.
5. Plastic vials.

2.4.3. Frozen Tissue

1. Plastic molds.
2. OCT embedding compound (Tissue Tek, Sakura)
3. Cotton swabs.
4. Dry ice.
5. 70% ethanol.

2.5. Immunostaining of Tissue Sections for Fluorescence Microscopy

1. Plastic staining box (microwaveable).
2. Slide warmer.
3. Histoclear (National Diagnostics #HS-200). Prepare three bottles with approximately 250 ml of Histoclear each. Label as Histoclear I, Histoclear II, and Histoclear III. Mark each use of these bottles on their labels. After the tenth use, discard Histoclear I, transfer the contents of Histoclear II into Histoclear I, and those of Histoclear III into Histoclear II. Refill the Histoclear III bottle from the stock container.
4. Ethanol. Prepare two bottles with 100% ethanol and label 100% Ethanol I and 100% Ethanol II. Use a maximum of ten times, transferring 100% Ethanol II contents into 100% Ethanol I and replenishing 100% Ethanol II contents from the stock ethanol container. Prepare also 95% and 70% ethanol bottles, and 50–50% mix of Histoclear and 100% ethanol and replenish from the stock every ten uses.
5. Distilled water.
6. Antigen unmasking solution (Vector #H3300). Prepare 500 ml (1:100 in water).
7. Staining rack (maximum 10 slides) and plastic coverplates (Thermo Fisher 7331017 and 72110017, respectively).
8. 0.5% Casein/0.5% Tween20/PBS. Put 0.5 g of casein in 100 ml of PBS. Stir for 1 h. Add 500 μl of Tween20 and stir for 20 more seconds.
9. Primary antibodies (Table 1).
10. Blocking buffer (Pierce 37515).
11. 0.1% Tween20/PBS.
12. Secondary antibodies (Table 2).
13. Aluminum foil.

Table 1
Primary antibodies and concentrations

Name	Source	Catalog number	Concentration
Rat anti-BrdU	Serotec	MCA2060	1:100
Rabbit anti-Ki67	Novocastra	NCL–Ki67p	1:800
Rabbit anti-PR	Dr. Shyamala Harris	N/A	1:800
Mouse anti-p63	Lab vision	MS-1081-P	1:200

Table 2
Secondary antibodies (all at 1:100)

Name	Source	Catalog number
Goat anti-rat FITC	Pierce	31629
Goat anti-rat Texas	Molecular Probes	T-6392
Goat anti-rabbit Texas	Molecular Probes	T-2767
Goat anti-mouse FITC	Pierce	31541

14. DAPI (Molecular Probes D-1306).
15. Vectashield (Vector H1000).
16. Coverglasses (Corning 125536).
17. Nail polish.

3. Methods

3.1. Bromodeoxyuridine and Pump Preparation

LRC identification is accomplished by labeling mammary epithelial cells with a thymidine analog during DNA synthesis (23, 24). Here we present a protocol using bromodeoxyuridine. BrdU is incorporated into the DNA of all the cells that undergo DNA replication during mammary gland development (3–6 weeks of age). Nearly all epithelial cells are labeled (24, 28). We use Brdu-filled, micro-osmotic pumps implanted intrascapularly vs. injections or administration in drinking water to obtain a constant rate of systemic BrdU (0.5 μl per hour) over a long time period (2 weeks). The protocol for BrdU dilution and pump filling follows.

3.1.1. BrdU Dilution

1. Dissolve 60 mg/ml BrdU in sterile PBS the day before the procedure (see Note 4).
2. Put BrdU powder in a small glass vial and add a magnet and sterile PBS. Close the vial. Put on a stirring plate, turn on heat (slowly, so that PBS does not evaporate, around 75°C) and stir. BrdU should go into dilution in approximately 5 min.
3. Aliquot (quickly!) in 1.5 ml tubes and freeze.

3.1.2. Filling the Pumps

1. Use hood under sterile conditions.
2. Warm up some PBS on a stirring plate (close to boiling point). If at any point the syringe used to fill the pumps gets clogged due to BrdU crystallization in the needle, put the needle in hot PBS for a couple of minutes.
3. Prepare paper towels, tube warmer, and vortex inside the hood.
4. Increase the temperature of the tube warmer up to 60–70°C, and put the BrdU tubes inside. After thawing the tubes, vortex and put back into the tube warmer.
5. Fill a 1 ml syringe (with a 27′-gauge needle already on) with the BrdU solution.
6. Weight pump and flow moderator together (~0.38 g).
7. Insert the syringe needle all the way into the pump from the top, while holding the pump vertically. Fill slowly until a small drop comes out.
8. Clean excess solution with a paper towel and insert flow moderator.
9. Weight pump and flow moderator ensemble. The pump should be at least 90% full (~0.47 g). If not, extract flow moderator, empty the pump with the needle and repeat the procedure.

3.2. Pump Implants

BrdU-filled pumps are implanted in 3-week-old females, at the beginning of pubertal mammary gland development. In order to reduce the invasiveness of the procedure, we implant smaller pumps (100 μl per pump) with half the desired delivery rate (0.25 μl per hour). Two of these pumps are implanted per animal. This allows labeling of the tissue for up to 2 weeks. Check with your institute's office of laboratory animal care for guidance before animal surgery.

1. Weigh the first mouse. Determine anesthetic dose (avertin, intraperitoneal, 0.2 ml/10 g body weight; or ketamine, intramuscular, in the thigh muscle, 0.02 ml/10 g body weight). Write down both weight and dose.

2. Anesthetize the mouse. Test for anesthetic by moderate pinch to one of the feet. There should be no response.
3. Tag the mouse if necessary, and write down the mouse id.
4. Put the mouse on the corkboard on its ventral side and strap down limbs with rubber bands, de-fur the surgery site if needed and swab with scrub or 70% ethanol.
5. Gently use tweezers to pick up the loose skin of the dorsal side and cut using surgical scissors, midway between tail and head. Make the incision wide enough to be able to introduce a pump.
6. Close scissors and use them to make a tunnel beneath the skin to free it from the body, but without tearing the skin or the body. Open the scissors slowly to widen the tunnel. Make enough room for two pumps.
7. Insert the two pumps with the flow moderator toward the posterior end of the mouse. After inserting the first pump, move it toward one side to make room for the next one.
8. Hold the two ends of the wound together with a forceps and use one (two at most) clip to close the wound (see Note 5).
9. Put the mouse on a paper towel in an empty, clean cage under the light of a lamp to keep it warm (you do not want to burn its skin, though!). Observe the mouse, it should wake up within 10–15 min.
10. Repeat all the steps above for the rest of the mice.
11. Check next day for losses.

3.3. Pump Explants

After tissue labeling, the pumps are removed and the animals are left to heal for several weeks. The surgical procedure is almost identical to that used for pump implantation.

1. Weight the first mouse. Determine anesthetic dose (avertin, intraperitoneal, 0.2 ml/10 g body weight; or ketamine, intramuscular, in the thigh muscle, 0.02 ml/10 g body weight). Write down both weight and dose.
2. Anesthetize the mouse. Test for anesthetic by moderate pinch to one of the feet. There should be no response.
3. Write down mouse ID, weight, and anesthetic dose.
4. Put the mouse on the corkboard on its ventral side and hold limbs with rubber bands and swab the surgery site.
5. Remove clips from implant procedure using clip remover.
6. Cut an incision on the dorsal side. Pumps tend to stick to the skin, so you may have to tear the skin (by pulling the pumps) in order to extract them. Alternatively, make this incision posterior with respect to the first one, where the flow moderators

of the pumps lie. This avoids having to cut through scar tissue, but will give an additional wound to the mouse.

7. Discard empty pumps.
8. Hold the two ends of the wound together with a forceps and use one (two at most) clip(s) to close the wound (see Note 5).
9. Put the mouse on a paper towel in an empty, clean cage under the light of a lamp to keep it warm and ensure recovery.
10. Check next day for losses.

3.4. Tissue Collection

The next period is called the *chase* during which, mammary gland development proceeds and proliferation of BrdU-labeled cells will halve the amount of BrdU in their nuclei. After three or four rounds of division, BrdU levels are marginal. Quiescent cells (24) or cells that can segregate their DNA differentially during asymmetric cell division (with a DNA strand constantly preserved in the mother cell (23)) retain the label over long periods of time. We use periods or *chases* of 13 weeks between pump extraction and tissue collection, although the results in terms of LRC frequency and distribution are similar to those obtained using 9-week-long chases.

In one set of mice, immediately after pump extraction at the end of the labeling period mice are euthanized, and mammary tissue is collected to control for the degree of labeling. These controls are extremely important, as they will determine the labeling efficiency of the pumps and the range of BrdU levels present in the samples. Approximately 80% of all the luminal epithelial cells should show some degree of labeling. Tissue is collected again at the end of the chase, when only LRC will remain labeled. We use the fourth pair of glands due to their larger size and ease of collection, which facilitates preservation of the morphology of the tissue. For nuclear feature analysis or quantification of clustering (*see* below), overnight formalin fixation is preferred, as it provides superior morphology. For colocalization studies with proteins that show weak staining, frozen samples produce less background, thus enhancing antigen detection via immunofluorescence.

3.4.1. Tissue Collection

1. Euthanize the mice using CO_2 followed by cervical dislocation, or as dictated by animal welfare guidelines.
2. For each mouse, place it in on its back over a corkboard. Stretch and restrain each pair of legs by pinning the feet to the corkboard and the extremities.
3. Spray the mouse with 70% ethanol and clean it with a paper towel.

4. Use sharp scissors to make a small cut in the midabdomen skin. Close the scissors and use them to tunnel beneath the skin, and then make an incision along the longitudinal axis, avoiding cuts to the peritoneal cavity.
5. Make two more cuts starting from the initial one and along the thigh muscle. Be careful not to section the femoral vein (see Note 6).
6. Gently pull the skin away from the peritoneum using forceps and teasing the skin from the peritoneal cavity with cotton swabs until the fourth mammary gland is exposed (see Note 7).
7. To remove the fourth mammary gland, place a piece of glassine paper (used for weighing) on top of the gland. Lift the mammary gland at the edge of the skin flap using forceps and make a small snip underneath to release it from the skin fascia. Cut all the attachments to the skin as you gently pull the gland away from it.

3.4.2. Formalin Fixation

1. Label all the cassettes necessary for your samples.
2. Place the sample on the piece of glassine paper and smooth it out as much as possible.
3. Insert the paper with the sample in a cassette.
4. Put the cassette in fresh 10% buffered formalin overnight (fixation).
5. Extract the sample from the cassette, peel off the paper and keep in a vial with 70% ethanol for later embedding and sectioning (see Note 8).

3.4.3. Frozen Tissue

1. Label all the molds necessary for your samples and fill one fourth with OCT.
2. Remove bubbles using the tip of a forceps or a cotton swab.
3. Collect each sample as described above, but do not use glassine paper. Snip the tissue gently away from the fascia. It will roll up on itself, but this is not a problem.
4. Prepare an ethanol bath in a glass dish with dry ice and 70% ethanol (poured on the dry ice).
5. Place tissue on top of the OCT layer, smoothing, centering and flattening it as much as possible.
6. After the ethanol bath "boils" for a bit, put molds with samples inside. Keep adding ethanol and dry ice at regular intervals, watching that ethanol does not get inside the molds.
7. When OCT freezes around the tissue, and before it freezes on the sample, fill the mold with OCT.
8. Repeat with all the samples in the animal under dissection (or with all the samples if only dissecting two or three mice).

9. Add dry ice/70% ethanol to the bath until all the OCT in all the molds is frozen and appears white (10–15 min).
10. Wrap the molds with foil and store at −80°C.

3.5. Immunostaining of Tissue Sections for Fluorescence Microscopy

Formalin-fixed samples preserved in 70% ethanol can then be embedded in paraffin and sectioned using a vibratome. We then use the immunofluorescence staining protocol described below to identify LRC in situ and to study them with respect to the expression of an additional marker, like Ki67 (proliferating cells), p63 (differentiated myoepithelial cells), or progesterone receptor (PR, mature luminal cells). Our protocol has been optimized for thin tissue sections (3–10 μm) and produces negligible cross-talk between imaging channels. If stored at −20°C, the slides stained following this protocol should be good for imaging for at least one month, although we have successfully used 6-month-old material. All our experiments include a no-primary antibody control and single antibody controls for dual staining that include each primary antibody with both secondary antibodies to ensure that there is no cross-reactivity to the nontarget primary.

1. Heat the slides in the slide warmer at 54–56°C for 30 min.
2. Deparaffinization. Wash slides in the plastic staining box with:
   a. Very warm Histoclear I – 10 min (microwave Histoclear for 3 min before using).
   b. Warm Histoclear II – 10 min (microwave for 1 min before using).
   c. Histoclear III – overnight.
   d. Histoclear 1/2 + EtOH 100% 1/2–5 min.
   e. 100% Ethanol I – 5 min.
   f. 100% Ethanol II – 3 min.
   g. 95% Ethanol – 3 min.
   h. 70% Ethanol – 3 min.
   i. Distilled water 3 × 5 min.
3. Antigen retrieval. Microwave at high temperature for 7 min ×3 and check the solution level between each time to prevent drying out the antigen retrieval.
4. Leave the slides cool down in solution for 20 min.
5. Wash with water for 5 min.
6. Transfer slides to staining rack with coverplate overlaid avoiding air bubble formation between the coverplate and the slide.
7. Wash with PBS.

8. Block the nonspecific background for 1 h at RT with the supernatant of a 0.5%Casein/0.1%Tween 20/PBS solution.
9. Incubate with the primary antibody against the additional marker diluted in blocking buffer (Table 1). This is an overnight incubation at 4°C.
10. Check for bubbles formed between the plastic surface and the slide.
11. Wash with 0.1% Tween 20/PBS.
12. Incubate with the secondary antibody against the additional marker diluted in PBS (Table 2). This incubation is for 1 h at RT. From this step on, everything is done *in the dark* (for example, by wrapping the rack with aluminum foil).
13. Wash with 0.1% Tween 20/PBS.
14. Block once again for non-specific background with 0.5% Casein/0.1% Tween 20/PBS for 1 h at RT.
15. Incubate with the primary BrdU antibody diluted in blocking buffer (Table 1). Incubate at 37°C for 1 h.
16. Wash with 0.1% Tween 20/PBS.
17. Incubate with the secondary antibody against the BrdU primary diluted in PBS (Table 2). Incubate for 1 h at RT.
18. Wash with 0.1% Tween 20/PBS.
19. Counterstain with DAPI diluted 1:1,000 in PBS for 10 min at RT.
20. Wash with 0.1% Tween 20/PBS.
21. Mount with Vectashield using a coverglass fixed with nail polish.
22. Store at −20°C.

3.6. In Situ Analysis

LRC studies allow for the in situ identification of a group of cells that contain a stem cell subpopulation (9, 18–20). Therefore, questions about ASC morphology and localization, difficult to investigate with other approaches, where cells are removed from their context in order to assess their "stemness," can be studied using LRC. We have developed a set of tools and strategies for in situ imaging and quantitative analysis of cell populations that we have integrated into a computational imaging platform called *R3D2* (26, 28, 29). In this section we describe several implementation details of R3D2 that are of interest for the analysis of mammary luminal LRC.

3.6.1. Imaging

Serial sections obtained from samples labeled, fixed and stained with the protocols described above (including no-primary antibody controls and cross-talk controls, if necessary) are imaged following a strategy that allows us to map the position of any given cell within the entire tissue. For practical reasons, only 4–10 sections per sample are used in each experiment (approximately 5,000–20,000 cells). As sections are cut parallel to the long axis of the tissue – the direction of ductal growth – all the types of ducts (primary, secondary and tertiary) should be present in each one of the slides.

1. Set the exposure time for low magnification acquisition (2.5×) of the nuclear counterstaining of one of the sections, and keep it constant for all the rest. Using times that produce bright but nonsaturated images should provide a good guide to the most appropriate values.
2. Take a low magnification image of the counterstaining all the sections. This can be accomplished by a precise sequence where an image is taken and the microscope stage, under the control of R3D2, is automatically moved to the next field of view. The sequence is then repeated until the entire slide is scanned. The focal plane should also be automatically adjusted through bilinear interpolation of the focal plane value determined manually at four different points (29). By tiling together images encompassing a single field of view into a mosaic one obtains an image of the entire tissue in each section. The resolution of this image is enough to identify ductal structures and the position of the lymph node, located at about one third of the length of the gland from the nipple, the site of convergence of the primary mammary ducts.
3. If an empty field was not included in your mosaic image, take a low magnification image of an empty field with the same wavelength and time used to acquire the mosaic. This background image can then be used to correct for uneven illumination.
4. Register or align the low magnification images. We automatically translate and rotate one image with respect to another in order to maximize the amount of overlap between epithelial structures, identified by brighter counterstaining (30) (see Note 9). Image registration allows efficient tracking of ducts across different sections.
5. Select ducts that can be tracked through all the sections under analysis. Chosen ducts should be homogeneously distributed in the stroma to include primary (central), secondary (intermediate), and tertiary (peripheral) ones.
6. Set the exposure time for high magnification acquisition (40×) of each channel. This exposure time should then be kept constant for all the images from the set of samples under study. One should first scan the samples to identify bright structures, like lymph nodes, and determine the best exposure time on those. For quantification purposes, it is

important that the images do not contain saturated pixels. Given the dynamic range of digital cameras, dim illumination is not a problem, but saturation renders the image useless for quantification of fluorescence intensity and can create artifacts. It is also wise to ensure that there is no bleed through from the red to the green channel or vice versa by imaging single antibody stained controls using the "other" wavelength.

7. Revisit the selected ducts using a high magnification lens. In R3D2, this is accomplished by clicking on the desired duct in the low magnification image and placing the right slide on the microscope stage. A color image including one channel per marker (DAPI, BrdU and an additional marker) is obtained by sequentially imaging while exciting the sample at 405 nm (i.e., DAPI), 488 nm (i.e., green fluorochrome), and 568 nm (i.e., red fluorochrome). The coordinates of the upper left corner of the high magnification image with respect to the upper left corner of the low magnification image are stored in the image header.
8. Take a high magnification color image of an empty field. This background image can then be used to correct for uneven illumination.
9. Map the high-magnification images onto the low-magnification ones. This can be automatically done if information about the coordinates of the upper left corner of each one of the images is stored during acquisition (see Note 10).

3.6.2. Preprocessing

Before images can be analyzed, they need to be corrected for background artifacts (31, 32). This is particularly necessary in the case of wide-field microscopy using fluorescent mercury lamps, which may provide uneven illumination in the field of view. Once the background has been corrected, the image is ready for quantification, assuming that no cross-talk occurs between channels (33) and that the imaging conditions were preserved between images. We then annotate our images interactively to determine the position of each cell, and to classify each one of the selected ducts as large, medium, or small (roughly equivalent to primary, secondary, or tertiary). Although there are automatic algorithms available for segmentation of cells in culture (34), their application to the in situ analysis of epithelial cells generally requires confocal imaging and extensive manual corrections (35, 36). This is due to the blurring effect of out-of-focus light on cell/nuclear outlines and the tight packing of epithelial cells in tissue sections, which causes large clusters of cells to be present in the images. Therefore, we opted for the labor-intensive, but potentially faster interactive annotation of the images.

3.6.2.1. Background Correction

1. For each one of the fields of view in the mosaic image, divide it by the background image. This is a pixel by pixel division, where the ratio of corresponding pixel values is the output.
2. Do a linear stretch of the pixel values of the resulting mosaic. This is accomplished with the following equation:

$$\text{output} = \frac{(\text{input} - \text{min\_in}) \times (\text{max\_out} - \text{min\_out})}{\text{max\_in} - \text{min\_in}} + \text{min\_out}, \quad (1)$$

where output is the resulting pixel value, input is the original one, min_in, max_in, min_out and max_out are, respectively, the minimum and maximum pixel values in the input and output images. Thus, for example, if the output image is an 8 bit image, where pixel values go from 0 to 255:

$$\text{output} = \frac{(\text{input} - \text{min\_in}) \times 255}{\text{max\_in} - \text{min\_in}}. \quad (2)$$

3.6.2.2. Interactive Annotation

1. In each one of the high magnification images, store the coordinates of a point close to the center of each one of the cell nuclei to analyze. We achieve this by clicking once on each nucleus and storing the coordinates of these marking in an annotation file (one per image).
2. Delineate the outline of the cells or nuclei that will be morphologically analyzed using the counterstaining image. We only use nuclei in focus, with clean, crisp outlines. The coordinates of each one of these outlines are also stored in the annotation file, linked to the appropriate cell.
3. As you move through the images, classify each duct as large, medium, or small and store this classification linked to the coordinates of the corresponding nuclei (e.g., in the same annotation file). We classify ducts according to these two criteria:
   a. Position within the fat pad. Large ducts tend to be central, while small ducts are peripheral (2). Based on the mapping of the high magnification image of the low magnification scan of the entire slide, it is possible to determine the exact position of each one of the ducts with respect to the border of the tissue.
   b. Thickness of periductal stroma. The stromal layer immediately around the ducts is thicker for large ducts than for small ones (28). We empirically determined that this layer has a thickness of up to 8.5 μm in small ducts, between 8.5 and 15.3 μm in medium ducts and larger than this in large ducts. To perform these measurements, tools to measure distances within an image are provided in R3D2.

3.6.2.3. Automatic Classification

1. Measure the staining intensity in each one of the channels for the nuclear markings as the average intensity (per channel) of a window centered at the nuclear marking. The window should be small enough to be contained within the nucleus, and large enough to account for potential variability of the staining. We have successfully used a squared window of 11×11 pixels to measure the intensity of each marker.
2. Calculate mean and standard deviation for the marking intensities in the no-primary antibody control for a given sample. Set intensity thresholds for each channel based on these values: frequently, the mean plus one or two standard deviations provide good separation between positive and negative measurements.
3. Apply the thresholds to the marking intensities measured on the sample images. Markings with intensity values higher than the threshold are classified as positive, while those with intensity values lower than the threshold are classified as negative.
4. Determine the accuracy of the classification by visually inspecting the results. If necessary, modify the threshold and reclassify the markings. It is important to note that the definition of the threshold for a certain channel should be maintained across samples. Thus, if the threshold is defined as the mean plus three standard deviations of the control marking intensities in one sample, this calculation should be repeated to obtain the threshold for that channel in the rest of the samples that we want to compare to this one.

3.6.3. Degree of Morphological Differentiation

Nuclear morphology can be studied using the nuclear outlines traced during the interactive annotation of high magnification images. Size, shape, and texture measurements are then used to characterize and compare different cell populations (28). In particular, nuclear size and chromatin texture distributions can be used to define the degree of morphological differentiation of a certain population, indicated by the frequency of cells with nuclear features of differentiation (28) (Table 3).

Table 3
Degree of morphological differentiation for several mammary epithelial populations

Cell population	Degree of morphological differentiation (%)
All LRC	32.3
Proximal LRC	30.6
Distal LRC	52.9
PR^+ LRC	69
Non LRC	68.2

1. For each delineated nucleus, measure size as the number of pixels contained within the nuclear outline.
2. For each delineated nucleus, measure chromatin texture as the standard deviation of the counterstaining (DAPI) intensities for all the pixels included within the nuclear outline (see Note 11).
3. Pooled together data from all available samples for the population of interest (e.g., LRC), as these measurements are independent of imaging conditions.
4. The threshold for morphological differentiation is given by:

$$y = \frac{1}{50}x, \tag{3}$$

with x, nuclear size, and y, chromatin texture values, respectively. This threshold was set based on chromatin texture vs. nuclear size plots comparing LRC and differentiated cells (PR^+). The degree of morphological differentiation for a certain population was defined as the percentage of cells in that population that falls below the threshold ($y < (1/50)\ x$). This percentage increases with the degree of differentiation of the population studied (Table 3).
5. Other morphological measurements can be used to evaluate shape. Shape factor (SF) and normalized bending energy (NBE) are two of them defined as:

$$\mathrm{SF} = \frac{4\pi A}{P^2}, \tag{4}$$

$$\mathrm{NBE} = \frac{\oint_P \kappa^2(p)dp}{P}, \tag{5}$$

where A is the area of the nucleus, P its perimeter, and $\kappa(p)$ the curvature of the nucleus at a certain point, p, on its perimeter. The shape factor measures the roundness of an object, while the normalized bending energy quantifies the irregularity of the perimeter. The distributions of these measurements for different populations can be compared using the Kolmogorov–Smirnov test (see Note 12).

3.6.4. Clustering

LRC and ASC have been shown to form clusters in several epithelial tissues, like the skin (5–7) or the small intestine (8, 9). We developed an automated algorithm to quantitatively determine the degree of clustering of a cell population and its statistical significance (26). Using this value, it is possible to compare different morphological structures to determine whether LRC clustering preferentially occurs in any one of them, as observed in the hair follicle bulge (7) or the bottom of the crypts of the small intestine (27). The algorithm is based on the construction of a graph that connects cell nuclei recapitulating the morphology of the tissue, thus allowing the identification of neighbor cells and the measurement of distances between cells using a standard unit, the graph edge; and on the *M* function that evaluates the fold-increase in clustering in the observed distribution with respect to a homogeneous distribution.

3.6.4.1. Relative Neighborhood Graph

1. Select the images corresponding to the morphological structure under analysis, e.g., large ducts.
2. In each image, build the Voronoi diagram of the nuclear markings that form it (26). This generates a partition of the image where each pixel is assigned to the closest nuclear marking.
3. Generate a Delaunay triangulation based on the Voronoi diagram. The nodes of this triangulation are the nuclear markings. Edges connect nuclear markings whose partitions in the Voronoi diagram shared an interface.
4. Build relative neighborhood graphs based on the Delaunay triangulations previously generated. In this graph, nodes correspond to nuclear markings, and edges connect the markings that correspond to neighbor cells. Given a set of points $\{p_1,\ldots,p_N\}$ representing the nuclear markings in an image, with N, the total number of markings, we process the edges of the Delaunay triangulation, so that an edge (p_i, p_j) is preserved if and only if p_i and p_j are at least as close to each other as they both are to any other node, i.e.,

$$d(p_i, p_j) \leq \max(d(p_i, p_k), d(p_j, p_k)), \tag{6}$$

$$\forall k = 1,\ldots,N, \quad k \neq i, j,$$

where $d(p_i, p_j)$ is the length of the edge between p_i and p_j. The distances between any two markings can be computed using Floyd's algorithm (37).
5. Refine the relative neighborhood graph by removing edges that are too large and forcings connections between points which are too close (as defined by their Euclidean distance) not to be neighbors. For this step we use two interactively defined thresholds which are the only parameters to the algorithm: edges longer than 1.05× the average edge length are deleted, and disconnected nuclei closer than 0.25× that same length are connected.

3.6.4.2. *M* Function Analysis

1. In each one of the images selected, choose the nuclear markings classified as positive for the population under study, *C*, e.g., BrdU in the case of LRC.
2. For each one of these positive nuclear markings evaluate the *M* function, given by:

$$M_i(r,C) = \frac{n_{iCr}}{n_{ir}} \Big/ \frac{N_C}{N}, \quad (7)$$

where r is the distance around marking p_i in C (measured on the relative neighborhood graph) where clustering will be evaluated; n_{iCr} is the number of neighbors of marking p_i belonging to population C within distance r; nir is the total number of neighbors of p_i (belonging to any population) within a distance r; NC is the total number of cells belonging to C in the image; and N is the total number of cells (in any population) in that same image. The numerator in (7) contains the fraction of neighbors that belong to C within distance r. The denominator is a ***benchmark***: the fraction of nuclei that belong to C in the entire area under study. Thus, clustering around a cell will result in $M_i(r, C) > 1$, with peaks at distances where the local density is the greatest. Regular patterns will have $M_i(r, C) < 1$. Finally, random distributions will have $M_i(r, C) = 1$. In general, $M_i(r, C) = k$ implies that the density of nuclei belonging to population C within distance r around p_i is k times that of the entire area under study.
3. Obtain the global M value for population C by averaging the M values for all the cells analyzed:

$$M(r,C) = \frac{\sum_{i=1}^{N_C} M_i(r,C)}{N_C}. \quad (8)$$

3.6.4.3. Significance Assessment

1. For each one of the images selected, scramble the position of the nuclear markings. This can be achieved by determining the number of positive (NC) and total (N) markings in each image and generating NC random numbers between 1 and N per image. If each nuclear marking has an index within its image, the markings corresponding to the randomly generated numbers are classified as positive for C, and the rest are classified as negative.
2. Repeat steps 7 and 8 above. This corresponds to obtaining the M function for a Monte Carlo simulation of a random distribution of positive nuclei markings.

3. Repeat steps 9 and 10 above as many times as simulations are necessary to attain the desired significance level. For a given P, this number can be computed as:

$$n_s = \frac{1}{P} - 1, \tag{9}$$

where n_s is the number of simulations necessary to obtain a confidence level of P. Thus, for example, for $P = 0.05$, $n_s = 19$. However, this approach is very sensitive to the presence of noise in the simulations, and tends to underestimate the significance of the results. For a more robust calculation, it is possible to discard the most extreme simulation results as random noise, thus requiring an increased number of simulations given by:

$$n_s = \frac{1 + n_o}{P} - 1, \tag{10}$$

where n_o is the number of simulations that will be discarded as outliers. We generally set $P = 0.05$ and $n_o = 9$ for a final $n_s = 199$.

4. For each distance, r, determine the maximum and minimum M values obtained from the simulations (after discarding the n_o most extreme values). This defines upper and lower significance thresholds $U(r, C)$ and $L(r, C)$.
5. Compare the values of $M(r, C)$ to $U(r, C)$ and $L(r, C)$ (Fig. 2). When $M(r, C) > U(r, C)$, the observed clustering at distance r is significant at significance level P, or in other words, the degree of clustering observed cannot be recapitulated in random distributions of population C. When $U(r, C) > M(r, C) > L(r, C)$, the observed clustering can be recapitulated in random distribution of C, and therefore, the observed distribution is not significantly different from random (not clustered). If $M(r, C) < L(r, C)$, the distribution of cells in C is not clustered, but regularly spaced (see Note 13).

3.6.5. Colocalization and Neighborhood

The algorithm described in Subheading 3.6.4 to quantify clustering within a cell population can be extended to determine the degree of association between the positions occupied by cells from two different populations, C_1 and C_2. The extended version of the M function used for this type of analysis can also be evaluated at multiple distances. By quantifying the degree of association of two populations, at distance $r = 0$, we obtain a measurement of their colocalization. Increasing this value to $r = 1, 2, \ldots$, allows us to identify the presence of different neighborhood relationships between different types of cells. This can be used to study the colocalization of LRC with proliferation markers or differentiation

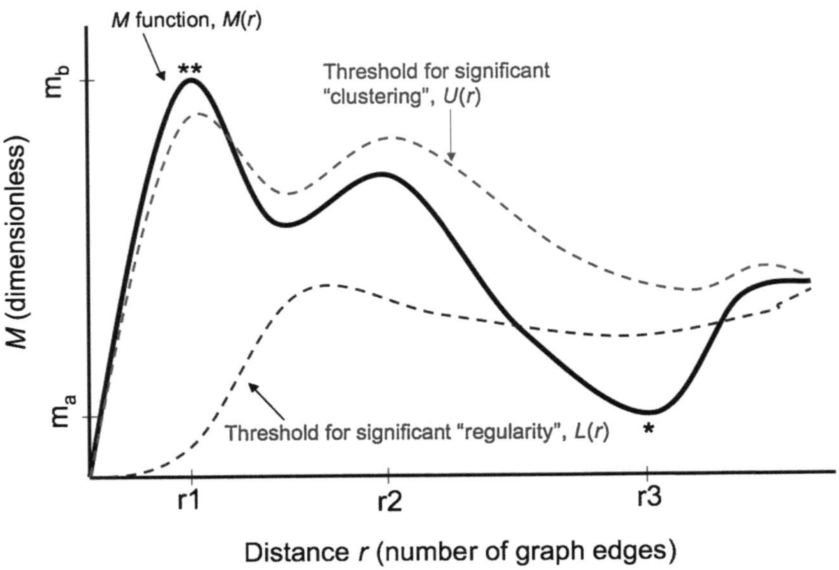

Fig. 2. *M* function interpretation. The cartoon represents a possible outcome of the clustering analysis described in the text. *M*, *U* and *L* functions are depicted. The double asterisk indicates a point where $M(r_1) > U(r_1)$. Therefore, at distance r_1 of any cell in the population under analysis, there is a density of other cells in that population that is m_b (>1) times that expected of a homogeneous distribution. The single asterisk indicates a point where $M(r_3) < L(r_3)$. Therefore, at distance r_3 of any cell in the population under analysis, there is a density of other cells in that population that is m_a (<1) times that expected of a homogeneous distribution

markers, like Ki67 or PR, respectively (28), or the association between putative niche markers and LRC.

1. Select the images corresponding to the morphological structure under analysis, e.g., large ducts.
2. Compute the relative neighborhood graph as described in Subheading 3.6.4.1.
3. To determine the distribution of cells in C_2 with respect to cells in C_1, choose the nuclear markings classified as positive for the reference population, C_1, in each one of the images selected.
4. For each one of these positive nuclear markings evaluate the *M* function, given by:

$$M_i(r, C_1, C_2) = \frac{n_{iC_2r}}{n_{ir}} \Big/ \frac{N_{C_2}}{N}, \tag{11}$$

where r is the distance around any cell in C_1 (measured on the relative neighborhood graph) where clustering will be evaluated; n_{iC_2r} is the number of neighbors of marking p_i belonging to population C_2 within distance r; N_{C1} is the total number of cells belonging to C_1 in the image; and N_{C2} is the total number of cells belonging to C_2 in that same image. The numerator in (11) contains the fraction of neighbors of a cell in C_1 that belong to C_2 within distance r. The denominator is a *benchmark*: the fraction of nuclei that belong to C_2 in the entire area under study. Thus, association between the localizations of C_1 and C_2 (e.g., immediate neighbors at $r=1$) around p_i will produce values of $M_i(r, C_1, C_2) > 1$. Patterns that tend to exclude each other (e.g., anti-colocalization at $r=0$) will have $M_i(r, C_1, C_2) < 1$. Finally, independent distributions will have $M_i(r, C_1, C_2) = 1$.

5. Obtain the global M value for the distribution of C_2 with respect to C_1 by averaging the M values for all the cells analyzed:

$$M(r,C_1,C_2) = \frac{\sum_{i=1}^{N_{C_1}} M_i(r,C_1,C_2)}{N_{C_1}}. \tag{12}$$

6. For each one of the images selected, scramble the position of the nuclear markings belonging to C_2.
7. Repeat steps 4 and 5 above. This corresponds to obtaining the M function for a Monte Carlo simulation of independent distributions of C_1 and C_2 nuclei.
8. Repeat steps 6 and 7 above as many times as simulations are necessary to attain the desired significance level, P. The number of simulations necessary for this can be computed as described in Subheading 3.6.4.1.
9. For each distance, r, determine the maximum and minimum M values obtained from the simulations (after discarding the n_0 most extreme values). This defines upper and lower significance thresholds $U(r, C_1, C_2)$ and $L(r, C_1, C_2)$.
10. Compare the values of $M(r, C_1, C_2)$ to $U(r, C_1, C_2)$ and $L(r, C_1, C_2)$. When $M(r, C_1, C_2) > U(r, C_1, C_2)$, the observed association at distance r is significant at significance level P, or in other words, the degree of association observed cannot be recapitulated by independent distributions. When $U(r, C_1, C_2) > M(r, C_1, C_2) > L(r, C_1, C_2)$, the observed association can be recapitulated by independent distributions, and therefore, the distribution of C_2 with respect to C_1 is not significantly different from an independent distribution. If $M(r, C_1, C_2) < L(r, C_1, C_2)$, cells in C_2 appear at distance r of cells in C_1 with a frequency that is smaller than expected if the distributions were independent.

3.6.6. Distribution Analysis

Once the images have been classified as representing one of the different morphological structures (or types of duct) present in the mammary gland, and having determined the positive or negative status of each one of the nuclear markings with respect to the different molecules under analysis (e.g., BrdU for LRC), it is possible to investigate the distribution of these different population across the different morphological structures present in the gland. However, we have observed that the range of LRC frequencies among different samples is quite large (from only 0.5% of all luminal epithelial cells up to 5%), and therefore it is necessary to normalize these frequencies before proceeding to investigate the distribution of LRC.

1. For each one of the n samples available, quantify the total frequency of nuclear markings positive for the population under analysis, C (e.g., LRC): $f_1, f_2, \ldots, f_n$.
2. For each one of the n samples, quantify the frequency of nuclear markings positive for C in each one of the m morphological structures under analysis: $f_{11}, f_{12}, \ldots, f_{1m}; \ldots; f_{n1}, f_{n2}, \ldots, f_{nm}$.
3. For each one of the samples, find the relative frequency of cells positive for C in each one of the morphological structures under analysis: $f_{11}/f_1, f_{12}/f_1, \ldots, f_{1m}/f_1; \ldots; f_{n1}/f_n, f_{n2}/f_n, \ldots, f_{nm}/f_n$.
4. Compare the following m vectors: $[f_{11}/f_1, f_{21}/f_2, \ldots, f_{n1}/f_n]$, $[f_{12}/f_1, f_{22}/f_2, \ldots, f_{n2}/f_n]$, …, $[f_{1m}/f_1, f_{2m}/f_2, \ldots, f_{nm}/f_n]$. Each one contains the relative frequencies in each sample for a certain morphological structure. Analysis of variance can be used to determine the probability that those vectors are sampled from distributions with the same variance (38). Based on that result, a t-test for samples with equal or different variances and Holmes correction for multiple groups (38) can be used to compare the means of each one of the vectors and determine whether significant differences exist in the distribution of the cells in C across different morphological structures (see Note 14).

4. Notes

1. These pumps come with flow moderators that will regulate release of the contents of the pump.
2. This is important since weight, and not age, determines the transition into puberty and the beginning of mammary gland development. The exact age/weight that ductal elongation

begins is strain and husbandry dependent, and may be modified in genetically engineered mice. If dealing with an unknown phenotype, prepare mammary gland wholemounts from 3-, 4-, and 6-week-old mice to establish the initiation and rate of ductal outgrowth.

3. Try and test this solution with two mice the day before the procedure. It is normally good for about 2–3 weeks. Then it loses anesthetic potential.
4. Prepare more than you would theoretically need (approximately twice as much) since some of it may be lost in clogged needles while filling the pumps.
5. Wound clips should stay on for 2 weeks.
6. The final incision looks like an inverted Y.
7. This gland runs all the way toward the dorsal side of the mouse.
8. Plastic vials are preferable, especially if the sample will be mailed for embedding and/or sectioning.
9. Alternatively, images can be manually registered by identifying corresponding points in consecutive images (29).
10. Otherwise, manual mapping by identifying corresponding points in high and low magnification images is also possible (29).
11. This measurement is independent of the imaging conditions (exposure time, lamp hours) unless there are saturated pixels

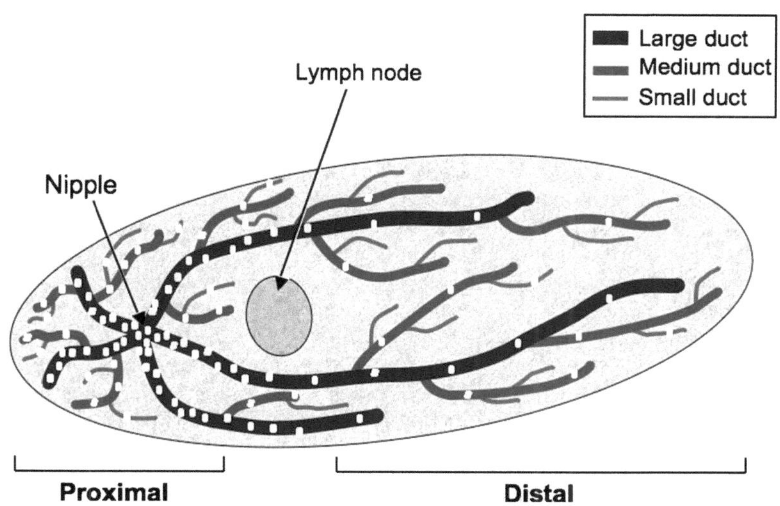

Fig. 3. LRC distribution in the mouse mammary gland. The cartoon shows the distribution of LRC in the luminal epithelium of the mouse mammary gland. LRC are more abundant in large ducts, proximal to the nipple, the site of origin of the gland. This suggests the presence of a proximal stem cell zone in mouse mammary tissue

in the counterstaining of the nuclei under analysis (in which case, the measurement is directly not applicable).

12. LRC present higher shape factors and lower normalized bending energies than non-LRC, indicating that their nuclei are more elliptical (28).
13. We found LRC to be clustered in large and medium ducts, and randomly distributed in small ducts (28).
14. Using this type of analysis, we found that the relative frequency of LRC is significantly higher in large and medium ducts than it is in small ducts. Furthermore, if ducts are classified as proximal or distal from the nipple, the site of origin of the gland, a 3.5-fold higher relative frequency of LRC is observed in the proximal end of the mammary gland (Fig. 3).

Acknowledgments

This work was supported by a predoctoral fellowship to RFG from the Department of Defense Breast Cancer Research Program (DAMD 17-03-1-0594), grants from the same institution to COS (DAMD 17-00-1-0227 and DAMD 17-00-1-0306), a grant to BEW from the National Cancer Institute (CA 8424306) and a grant to MHBH funded by the National Institute of Environmental Health Sciences and the National Cancer Institute (U01 ES012801).

References

1. Chu EY, Hens J, Andl T, Kairo A, Yamaguchi TP, Brisken C, Glick A, Wysolmerski JJ, Millar SE (2004) Canonical WNT signaling promotes mammary placode development and is essential for initiation of mammary gland morphogenesis. Development 131: 4819–4829
2. Hennighausen L, Robinson GW (2001) Signaling pathways in mammary gland development. Dev Cell 1:467–475
3. Smalley M, Ashworth A (2003) Stem cells and breast cancer: a field in transit. Nat Rev Cancer 3:832–844
4. Adriance MC, Inman JL, Petersen OW, Bissell MJ (2005) Myoepithelial cells: good fences make good neighbors. Breast Cancer Res 7:190–197
5. Savill NJ, Sherratt JA (2003) Control of epidermal stem cell clusters by Notch-mediated lateral induction. Dev Biol 258:141–153
6. Morris RJ, Liu YP, Marles L, Yang ZX, Trempus C, Li SL, Lin JS, Sawicki JA, Cotsarelis G (2004) Capturing and profiling adult hair follicle stem cells. Nat Biotechnol 22:411–417
7. Tumbar T, Guasch G, Greco V, Blanpain C, Lowry WE, Rendl M, Fuchs E (2004) Defining the epithelial stem cell niche in skin. Science 303:359–363
8. Potten CS, Loeffler M (1990) Stem cells: attributes, cycles, spirals, pitfalls and uncertainties. Lessons for and from the crypt. Development 110:1001–1020
9. Potten CS, Booth C, Tudor GL, Booth D, Brady G, Hurley P, Ashton G, Clarke R, Sakakibara S, Okano H (2003) Identification of a putative intestinal stem cell and early lineage marker; musashi-1. Differentiation 71:28–41
10. Spradling A, Drummond-Barbosa D, Kai T (2001) Stem cells find their niche. Nature 414:98–104

11. Shackleton M, Vaillant F, Simpson KJ, Stingl J, Smyth GK, Asselin-Labat ML, Wu L, Lindeman GJ, Visvader JE (2006) Generation of a functional mammary gland from a single stem cell. Nature 439:84–88
12. Stingl J, Eirew P, Ricketson I, Shackleton M, Vaillant F, Choi D, Li HYI, Eaves CJ (2006) Purification and unique properties of mammary epithelial stem cells. Nature 439:993–997
13. Novaro V, Roskelley CD, Bissell MJ (2003) Collagen-IV and laminin-1 regulate estrogen receptor alpha expression and function in mouse mammary epithelial cells. J Cell Sci 116:2975–2986
14. Liu BY, McDermott SP, Khwaja SS, Alexander CM (2004) The transforming activity of Wnt effectors correlates with their ability to induce the accumulation of mammary progenitor cells. Proc Natl Acad Sci USA 101: 4158–4163
15. Asselin-Labat ML, Shackleton M, Stingl J, Vaillant F, Forrest NC, Eaves CJ, Visvader JE, Lindeman GJ (2006) Steroid hormone receptor status of mouse mammary stem cells. J Natl Cancer Inst 98:1011–1014
16. Braun KM, Watt FM (2004) Epidermal label-retaining cells: Background and recent applications. J Invest Dermatol Symp Proc 9:196–201
17. Tsujimura A, Koikawa Y, Salm S, Takao T, Coetzee S, Moscatelli D, Shapiro E, Lepor H, Sun TT, Wilson EL (2002) Proximal location of mouse prostate epithelial stem cells: a model of prostatic homeostasis. J Cell Biol 157:1257–1265
18. Burger PE, Xiong X, Coetzee S, Salm SN, Moscatelli D, Goto K, Wilson EL (2005) Sca-1 expression identifies stem cells in the proximal region of prostatic ducts with high capacity to reconstitute prostatic tissue. Proc Natl Acad Sci USA 102:7180–7185
19. Oshima H, Rochat A, Kedzia C, Kobayashi K, Barrandon Y (2001) Morphogenesis and renewal of hair follicles from adult multipotent stem cells. Cell 104:233–245
20. Potten CS (2004) Keratinocyte stem cells, label-retaining cells and possible genome protection mechanisms. J Investig Dermatol Symp Proc 9:183–195
21. Asselin-Labat ML, Sutherland KD, Barker H, Thomas R, Shackleton M, Forrest NC, Hartley L, Robb L, Grosveld FG, van der Wees J, Lindeman GJ, Visvader JE (2007) Gata-3 is an essential regulator of mammary-gland morphogenesis and luminal-cell differentiation. Nat Cell Biol 9:201–209
22. Clayton E, Doupe DP, Klein AM, Winton DJ, Simons BD, Jones PH (2007) A single type of progenitor cell maintains normal epidermis. Nature 446:185–189
23. Smith GH (2005) Label-retaining epithelial cells in mouse mammary gland divide asymmetrically and retain their template DNA strands. Development 132:681–687
24. Welm BE, Tepera SB, Venezia T, Graubert TA, Rosen JM, Goodell MA (2002) Sca-1(pos) cells in the mouse mammary gland represent an enriched progenitor cell population. Dev Biol 245:42–56
25. Blanpain C, Horsley V, Fuchs E (2007) Epithelial stem cells: turning over new leaves. Cell 128:445–458
26. Fernandez-Gonzalez R, Barcellos-Hoff MH, Ortiz-de-Solorzano C (2005) A tool for the quantitative spatial analysis of complex cellular systems. IEEE T Image Process 14: 1300–1313
27. Potten CS, Owen G, Booth D (2002) Intestinal stem cells protect their genome by selective segregation of template DNA strands. J Cell Sci 115:2381–2388
28. Fernandez-Gonzalez R (2006) In: Bioengineering, Vol. PhD, UC Berkeley/UC San Francisco, Berkeley
29. Fernandez-Gonzalez R, Jones A, Garcia-Rodriguez E, Chen PY, Idica A, Lockett SJ, Barcellos-Hoff MH, Ortiz de Solorzano C (2002) System for combined three-dimensional morphological and molecular analysis of thick tissue specimens. Microsc Res Tech 59:522–530
30. Arganda-Carreras I, Fernandez-Gonzalez R, Ortiz de Solorzano C (2004) In: Twentysixth annual international conference of the engineering in medicine and biology society, vol 1. San Francisco, CA, pp 1691–1694
31. Fernandez-Gonzalez R, Deschamps T, Idica A, Malladi R, Ortiz de Solorzano C (2004) Automatic segmentation of histological structure in mammary gland tissue sections. J Biomed Opt 9:444–453
32. Chamberlain CE, Kraynov VS, Hahn KM (2000) Imaging spatiotemporal dynamics of Rac activation in vivo with FLAIR. Method Enzymol 325:389–400
33. Zimmermann T (2005) Spectral imaging and linear unmixing in light microscopy. Adv Biochem Eng Biotechnol 95:245–265
34. Fernandez-Gonzalez R, Barcellos-Hoff MH, Ortiz de Solorzano C (2004) Quantitative image analysis in mammary gland biology. J Mammary Gland Biol Neoplasia 9:343–359

35. Ortiz de Solorzano C, Rodriguez EG, Jones A, Pinkel D, Gray JW, Sudar D, Lockett SJ (1999) Segmentation of confocal microscope images of cell nuclei in thick tissue sections. J Microsc 193:212–226
36. Wahlby C, Sintorn IM, Erlandsson F, Borgefors G, Bengtsson E (2004) Combining intensity, edge and shape information for 2D and 3D segmentation of cell nuclei in tissue sections. J Microsc 215:67–76
37. Sedgewick R (2003) Algorithms in Java, part 5: graph algorithms, Addison Wesley Professional
38. Glantz SA (2005) Primer of biostatistics, McGraw-Hill Medical, New York

Chapter 2

Limiting-Dilution Transplantation Assays in Mammary Stem Cell Studies

Irineu Illa-Bochaca, Rodrigo Fernandez-Gonzalez, Dawne N. Shelton, Bryan E. Welm, Carlos Ortiz-de-Solorzano, and Mary Helen Barcellos-Hoff

Abstract

Mammary reconstitution assays can be used to measure the stem cell frequency within an epithelial population by transplanting increasingly diluted single-cell preparations of the population of interest. There are fundamental steps in the single-cell isolation protocol which are directly related to the number of single epithelial cells obtained. Once single-cell suspensions have been obtained, serial dilutions are prepared and transplanted into the cleared fat pads of the host mice. After 8–10 weeks, the transplanted fat pads are reevaluated for the presence of epithelial outgrowths. Based on the frequency of no outgrowth for each one of the transplanted dilutions, it is possible to estimate the frequency of mammary repopulating cells present in a given cell population. Here, we give details on how to carry out all these steps.

Key words: Cell isolation, Differential centrifugation, Cleared fat-pad, Transplantation, Outgrowth, Wholemount

1. Introduction

The most important data suggesting the presence of mammary-specific stem cells were obtained in a series of break-through experiments carried out in the 1950s in the Cancer Research Lab, at the University of California, Berkeley. When mouse mammary epithelial cells were transplanted into cleared fat pads in which the epithelial compartment had been excised (1), the transplanted cells were able to proliferate and undergo ductal morphogenesis that filled the fat pad, and upon pregnancy, the resulting outgrowths produced milk. Similar results were observed when

Irina M. Conboy et al. (eds.), *Protocols for Adult Stem Cells*, Methods in Molecular Biology, vol. 621,
DOI 10.1007/978-1-60761-063-2_2, © Springer Science+Business Media, LLC 2010

tissue fragments, rather than isolated epithelial cells, were used for transplantation. The age of the donor did not have an effect on the outgrowth potential of the transplant (2). These experiments clearly show that there must be a pool of cells within mammary epithelium with extensive regenerative potential. Furthermore, the fact that serial transplantation experiments can go on up to five or six iterations with similar efficiency (3) indicates that these cells not only can give rise to a fully functional mammary gland, but are also able to self-renew. Similar properties have later been shown for rats (4–6) and for human mammary epithelial cells (transplanted into mouse fat pads) (7, 8).

Here, we describe the materials and methods necessary to carry out transplantation experiments using mammary tissue. Then, we describe the statistical methods necessary to assay the fraction of mammary-repopulating cells within a certain population based on the frequency of outgrowth formation for different numbers of transplanted cells. These methods are generally applicable to any type of limiting-dilution transplantation experiments, the gold standard to assess the regenerative potential of cell populations. We close the chapter with an appendix containing the code of Matlab (Mathworks, Natick, MA) routines that implement the above-mentioned statistical methods.

2. Materials

2.1. Mouse Mammary Epithelial Cell Isolation

2.1.1. Mammary Gland Collection

1. Collagenase/hyaluronidase (Stem Cell Technologies #07912).
2. Epicult medium-B (Stem Cell Technologies #05602).
3. Epicult supplement (Stem Cell Technologies).
4. Fetal Bovine Serum (FBS).
5. Warm water bath.
6. CO_2 source.
7. Surgery instrumentation: scissors and two forceps.
8. Corkboard and rubber bands.
9. DMEM/F12 (Gibco).
10. Sterile scalpel.
11. Sterile 150 mm Petri dish.
12. Environmental shaker.

2.1.2. Differential Centrifugation

1. Centrifuge (preferable with fast brake).
2. 50 ml centrifuge tubes.
3. Hank's solution or HBSS.

2.1.3. Single Cell Isolation

1. Trypsin–EDTA.
2. Dispase (Stem Cell Technologies, 07913).
3. DNAseI (Sigma, D4263).
4. Ammonium chloride.
5. Trypan blue dye.
6. 40 μm cell strainer (BD Falcon).
7. Microscope.
8. Hemocytometer.

2.2. Epithelial Cell Transplantation

2.2.1. Fat Pad Clearing

1. Avertin (ketamine can also be used as anesthetic).
   (a) Stock solution: dissolve 25 g of 2,2,2-tribromoethanol (Sigma T48402 100 g) into 15.5 ml tert-amyl alcohol (Fisher A730-1). For this, you will have to stir for 12 h in a foil-wrapped bottle (avertin is light-sensitive). Wash the magnet that you will use with hot water before starting. Store at room temperature.
   (b) 0.9% saline: 0.9 g of NaCl in 100 ml deionized or distilled water. Filter sterilize with a 0.2 μm (or 0.22 μm) filter.
   (c) Working solution: stir 0.5 ml stock solution into 39.5 ml of 0.9% saline for 12 h. Wash the magnet that you will use for this with hot water before starting to stir. Store at 4°C (see Note 1).
   (d) Dosage: avertin, 0.2 ml/10 g body weight; ketamine, 0.02 ml/10 g body weight.
2. Corkboard (with rubber bands, no pins!).
3. 70% ethanol.
4. Paper towels.
5. Surgery instrumentation: scissors and two forceps.
6. Cotton swabs.
7. Thermocautherizer (Max Wax, Pen500).

2.2.2. Mammary Epithelial Cell Transplantation

1. Hamilton syringe (Hamilton #705-N).
2. Wound clips, clipping machine and clip remover.
3. Extra clean cage.
4. Lamp.

2.3. Outgrowth Collection and Staining

2.3.1. Wholemount Collection

1. CO_2 source.
2. Mice treated as described in Subheadings 3.2.1 and 3.2.2.
3. Corkboard and rubber bands.
4. 70% ethanol.
5. Paper towels.

6. Surgery instrumentation: scissors and two forceps.
7. Cotton swabs.
8. Waxed or glassine paper.

2.3.2. Fixation

1. Carnoy's fluid.

2.3.3. Staining

1. Alcohol 70%.
2. Toluene (JT Baker #9460-01).
3. Glass vials.
4. Methyl salicylate (Sigma #M6752).

3. Methods

3.1. Mouse Mammary Epithelial Cell Isolation

Limiting-dilution transplantation assays can be used to compare the regenerative potential of multiple populations. The first step of this procedure is to isolate mammary epithelial cells. When performing epithelial cell isolation it is essential to avoid contamination with blood cells. Differential centrifugation is the critical step to eliminate this type of contamination. During tissue disaggregation, at the point at which epithelial cells remain bound in organoids, most blood cells and fibroblasts can be eliminated using differential centrifugation. A final treatment with ammonium chloride is used to lyse the remaining erythrocytes. All the isolation process should be carried out in a laminar flow hood to avoid possible contamination. After isolation, cells are kept on ice until transplantation, which should be done immediately.

3.1.1. Dissection

1. Mix one part of collagenase/hyaluronidase with nine parts of Epicult medium and 5% FBS. Use a water bath to prewarm at 37°C.
2. Euthanize the mice by CO_2 inhalation followed by cervical dislocation or as indicated by the institutional animal welfare committee.
3. Dissect the mice and extract both fourth mammary glands as described in Chapter 1, Subheading 3.4.
4. Collect the portions of the tissue whose regenerative potential is going to be assessed (e.g., proximal vs. distal) in DMEM/F12 (Gibco) supplemented with 10% FBS. Maintain the tubes on ice as tissues are collected.
5. Mince tissue with a sterile scalpel on a 150 mm sterile Petri dish for about 3 min, until a homogenous paste is obtained.
6. Transfer minced tissue into tubes filled with collagenase/ hyaluronidase and incubate under constant agitation in an

environmental shaker for 1 h at 37°C. Use approximately 5 ml of C/H solution for each gram of tissue.

7. Spin the tubes at 450 g for 5 min.
8. Discard the supernatant. Resuspend organoids in cold HBSS supplemented with 2% FBS. From now on, all pipettes should be precoated with this solution.

3.1.2. Differential Centrifugation

1. After resuspending the organoids, do pulse centrifugation by accelerating the centrifuge to 450 g, and then stopping it with fast brake on.
2. Collect the supernatant, which contains the contaminating cells, and examine a drop on a small Petri dish to ensure that organoids are not excluded with the supernatant.
3. Repeat steps 7 and 8 up to five times (see Note 2).

3.1.3. Single Cell Isolation

1. Add 2 ml prewarmed Trypsine/EDTA (Stemcell Technology) to each tube.
2. Transfer the pellet (at the bottom of the tube) into a 6 well plate and pipet up and down with a 1,000 µl tip for 1–3 min (see Note 3).
3. Add about 10 ml of cold HBSS containing 2%FBS and 2 ml of Trypsin. Transfer to centrifuge tubes and spin again at 450 g for 5 min.
4. Remove the supernatant carefully because some of the cells could remain as a stringy mass floating in the supernatant. Leave some fluid in the tube.
5. Add 5 ml of prewarmed dispase containing 0.1 mg/ml DNAse per tube. Mix the content of the tubes up and down for at least 1 min.
6. Add 5 ml HBSS/FBS, and then spin down at 45 g for 5 min.
7. Remove the supernatant and resuspend the pellet with 4 ml per tube with one part of HBSS/FBS and four parts of ammonium chloride to lyse remaining erythrocytes.
8. Centrifuge at 450 g for 5 min.
9. Remove supernatant and add 5 ml DMEM/F12 medium.
10. Filter suspension through a 40 µm mesh (Falcon tube insert) into a new 14 ml tube. Place the filter in a lateral position and turn it when the current corner gets clogged.
11. Centrifuge again at 450 g for 5 min and discard the supernatant. Add 1 ml of DMEM/F12.
12. Measure the concentration of cells in a hemocytometer and prepare the correspondent dilutions so that 10-µl injections can be used for transplantation.

3.2. Epithelial Cell Transplantation

Before mammary cell transplantation, it is necessary to remove the epithelium from the glands of the host mice. These should be 3-week-old mice prior to starting puberty. After clearing the fat pads, cells are transplanted via injection of 10 μl of a cell suspension with a specific concentration of cells. At least three different concentrations are necessary to assess the regenerative potential of a certain cell population, although more groups result in more accurate estimates. Use at least two or three groups around the predicted mammary-repopulating cell frequency in the population under study. For example, if 1/100 cells is expected to be mammary-repopulating, then use 1,000 control, in which all transplants should be successful, 200, 100, and 50. The more groups included between the 200 and the 50-cell ones, the best estimate will be obtained in this example. Include at least 20 fat pads per group, although this number can be computed more rigorously using power analysis (9).

3.2.1. Clearing Fat Pads

Check with your institute's office of laboratory animal care for guidance before animal surgery.

1. Weigh the first mouse. Determine anesthetic dose (avertin, intraperitoneal, 0.2 ml/10 g body weight; or ketamine, intramuscular – in the thigh muscle – 0.02 ml/10 g body weight). Write down both weight and dose.
2. Anesthetize the mouse. Test for anesthetic by moderate pinch to one of the feet. There should be no response.
3. Place the anesthetized mouse on its back over a corkboard. Stretch and restrain each pair of legs with a rubber band that runs around the corkboard and the extremities.
4. De-fur the surgery site if needed and spray the mouse abdomen with 70% alcohol and clean it with a sterile napkin, or use a surgical scrub.
5. Use sharp scissors to make a small cut in the midabdomen skin after lifting the loose skin with forceps. Close the scissors and use them to tunnel through the skin, and then make an incision along the longitudinal axis. Avoid cutting the peritoneal cavity.
6. Make two more incisions from the initial one and following the thigh muscle. Be careful not to section the femoral vein (see Note 4).
7. Use tweezers and cotton swabs to pull the skin gently away from the peritoneal cavity and expose the fourth mammary gland (see Note 5).
8. Examine the mammary gland with high intensity illumination. The lymph node is visible as a pearly bump the size of a bean. A large vessel runs the length of the gland and bifurcates

near the nipple. With practice and oblique illumination, it is possible to make out the nipple.

9. Lift the mammary gland at the caudal edge using forceps and cut all the attachments to the skin as you gently pull the gland away from it until the caudal third below the lymph node of the gland has been freed.
10. Using a thermocautherizer, remove the portion of the mammary gland below the lymph node (including it). At least 2/3 of the mammary gland must remain intact and in place.
11. Use a thermocautherizer to burn off the area where the nipple was located. This will ensure that the dissected region does not contain any residual epithelial cells.
12. Also cut the connection between the fourth and fifth mammary glands using the thermocautherizer.

3.2.2. Mammary Epithelial Cell Transplantation

1. Use a Hamilton syringe to inject 10 μl of the mammary epithelial cells previously obtained (see Subheading 3.2.1).
2. Place the skin in its original position, over the abdomen, and repeat the same procedure (steps 6–11) for the other side of the animal.
3. Hold the two ends of the wound together with a forceps and use three or four clips to close it (see Note 6).
4. Put the mouse on a paper towel in an empty, clean cage under the light of a lamp to keep it warm (you do not want to burn its skin, though!).

3.3. Outgrowth Collection and Staining

To evaluate the repopulation potential of the transplanted epithelial cells, the outgrowth frequency was measured 6 weeks after transplantation. Prior to this, it is necessary to collect, fix, and stain the mammary glands.

3.3.1. Mammary Gland Collection

1. Euthanize the mice using CO_2 followed by cervical dislocation or as indicated by the institution animal welfare committee.
2. For each mouse, place it in on its back over a corkboard. Stretch and restrain each pair of legs by pushing pins into the corkboard and the extremities.
3. Spray the mouse with 70% ethanol and clean it with a paper towel.
4. Use sharp scissors to make a small cut in the midabdomen skin. Close the scissors and use them to tunnel through the skin, and then make an incision along the longitudinal axis, avoiding cuts to the peritoneal cavity.

5. Make two more cuts starting from the initial one and along the thigh muscle. Be careful not to section the femoral vein (see Note 4).
6. Gently pull the skin away from the peritoneum using forceps and cotton swabs until the fourth mammary gland is exposed (see Note 5).
7. Lift the mammary gland at the location of the nipple using forceps and cut all the attachments to the skin as you gently pull the gland away from it.
8. Place a piece of waxed or glassine paper over the gland and press softly with a cotton swab (see Note 7).
9. Use the tweezers to lift the mammary gland and the paper together. Use the scissors to cut the remaining connections between the mammary gland and the tissue.

3.3.2. Fixation

Submerge the gland in Carnoy's fluid and leave overnight.

3.3.3. Staining

1. After fixation, place the tissue in 70% alcohol for 15 min, followed by 5 min in running tap water.
2. Put the wholemounts in carmine alum for 24 h.
3. Use ethanol in increasing graduation (70, 90, and 95%) to dehydrate the tissue. For an optimal dehydration tissue must be 15 min in each ethanol wash.
4. Clear the wholemount in toluene at least for 15 min.
5. To store the glands, peel off the paper and place them in a glass vial containing methyl salicylate.
6. To examine a gland under the dissecting microscope or to image it, pour some methyl salicylate in a Petri dish and put the gland inside, covered with a coverslip.
7. Outgrowths are scored positive if they fill at least 10% of the fat pad (Figs. 1a, b). The character of the outgrowth can be assessed by enumerating branches per length of duct, presence of end buds, and degree of alveolar differentiation.

3.4. Data Analysis

The frequency of no outgrowth formation is obtained for each one of the cell concentrations used after scoring the transplants. Analysis is then done using a single-hit Poisson model (10) to estimate the frequency of mammary-repopulating cells present in the initial population. Briefly, this model assumes that a single stem cell is enough to give rise to an outgrowth, and that the injected cells are sampled from a homogeneously suspended population. The frequency of outgrowth-forming cells and 95% confidence intervals are estimated by fitting the experimental data (frequency of no outgrowth per cell injection size) to a Poisson model using χ^2 minimization. A set of Matlab (Mathworks,

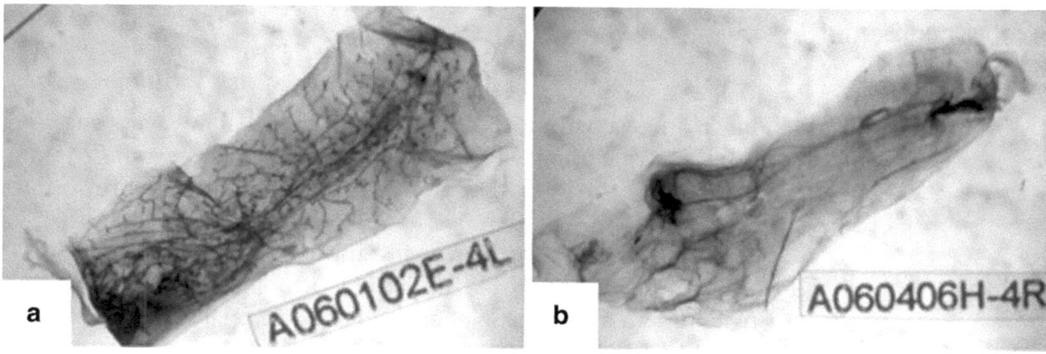

Fig. 1. Outgrowth scoring. (**a**) Wholemount preparation showing an outgrowth that filled the host fat pad. (**b**) Wholemount preparation showing a fat pad where no outgrowth formed

Natick, MA) functions is provided in the appendix to do this calculation.

1. Estimate the frequency of mammary-repopulating cells and 95% confidence intervals for each one of groups under analysis.
2. Compare the estimates for each group. Nonoverlapping 95% confidence intervals indicate a significant difference between the groups (11) (see Note 8).

4. Notes

1. Try and test this solution with two mice the day before the procedure. It is normally good for about 2–3weeks. Then it loses anesthetic potential.
2. The pellet, at the bottom of the tubes, should contain mainly organoids; single cells, fibroblast, and blood cells are discarded with the supernatant.
3. The sample will become stringy due to lysis of cells and the consequent DNA release.
4. The final incision looks like an inverted Y.
5. This gland runs all the way toward the dorsal side of the mouse.
6. Wound clips should stay on for 2 weeks.
7. The gland will adhere to the waxed paper and will remain stretched during the fixation and staining processes, which allows for better observation of the ductal morphology.
8. Using this approach, we found a significant 2.5-fold increase in the frequency of mammary-repopulating cells in proximal vs. distal mammary epithelium (12).

Acknowledgments

This work was supported by a predoctoral fellowship to RFG from the Department of Defense Breast Cancer Research Program (DAMD 17-03-1-0594), grants from the same institution to COS (DAMD 17-00-1-0227 and DAMD 17-00-1-0306), a grant to BEW from the National Cancer Institute (CA 8424306), and a grant to MHBH funded by the National Institute of Environmental Health Sciences and the National Cancer Institute (U01 ES012801).

Appendix I

fMC.m

```
% [estimateMC, estimateML, estimateWM] = fMC(I, r, n, x, k, estimate)
% Estimates the frequency of "hit-cells" in a limiting dilution
experiment (minimum chi-squared method, see [1]).
%
% I = number of cell doses or groups.
% r = number of negative outcomes in each group.
% n = number of repeats in each group.
% x = number of cells used in each group.
% k (optional) = determines the confidence interval to be computed.
This value corresponds to the value of
%     Student's t distribution with n-1 degrees of freedom at the
desired confidence level divided by two
%     (for large samples, a Gaussian). Thus, for example, the 95%
confidence interval can be computed with
%     k = 1.96 (default value).
% estimate (optional) = an estimate for the value to be computed. If
this parameter is not provided, an estimate
%    is calculated using maximum likelihood (fML). To provide an
external estimate, provide also the k value.
%
% estimateMC = estimate of hit cell frequency using minimum chi-
squared.
% estimateML = estimate of hit cell frequency using maximum likelihood
or the user provided estimate.
% estimateWM = estimate of hit cell frequency using weighted mean (if
maximum likelihood was used to estimate estimateML).
%
% [1] Taswell C., "Limiting dilution analysis for the determination of
immunocompetent cell frequencies", The Journal of Immunology
%     126(4):1614 (1981).
%
% Rodrigo Fernandez-Gonzalez
% 06/28/2006

function varargout = fMC(varargin)

switch (nargin)
      case 4
            % Number of experimental groups.
            I = varargin{1, 1};
            if (prod(size(I)) ~= 1 || (~ isnumeric(I)))
                  error('The first parameter has to be a number.');
            end

            % Number of fat pads with no outgrowth per group.
            r = varargin{1, 2};
            if (prod(size(r)) ~= I)
                  error('The size of the vector parameters is given by
the first argument.');

            else
                  r = reshape(r, 1, I);
            end
```

```
            % Total number of fat pads per group.
            n = varargin{1, 3};
            if (prod(size(n)) ~= I)
                  error('The size of the vector parameters is given by
the first argument.');
            else
                  n = reshape(n, 1, I);
            end

            % Number of cell transplanted per group.
            x = varargin{1, 4};
            if (prod(size(x)) ~= I)
                  error('The size of the vector parameters is given by
the first argument.');
            else
                  x = reshape(x, 1, I);
            end

            % By default the confidence interval to compute is 95%.
            k = 1.96;

            % Obtain an estimate of the frequency that is to be
computed.
            [estimateML estimateWM] = fML(I, r, n, x);


      case 5
            % Number of experimental groups.
            I = varargin{1, 1};
            if (prod(size(I)) ~= 1 || (~ isnumeric(I)))
                  error('The first parameter has to be a number.');
            end

            % Number of fat pads with no outgrowth per group.
            r = varargin{1, 2};
            if (prod(size(r)) ~= I)
                  error('The size of the vector parameters is given by
the first argument.');
            else
                  r = reshape(r, 1, I);
            end

            % Total number of fat pads per group.
            n = varargin{1, 3};
            if (prod(size(n)) ~= I)
                  error('The size of the vector parameters is given by
the first argument.');
            else
                  n = reshape(n, 1, I);
            end

            % Number of cell transplanted per group.
            x = varargin{1, 4};
            if (prod(size(x)) ~= I)
                  error('The size of the vector parameters is given by
the first argument.');
            else
```

```
            x = reshape(x, 1, I);
        end

        % Confidence interval.
        k = cell2struct(varargin(1, 5), 'k', 1);
        k = k.k;
        if (prod(size(k)) ~= 1 || (~ isnumeric(k)))
            error('The fifth parameter has to be a number.');
        end

        % Obtain an estimate of the frequency that is to be
computed.
        [estimateML estimateWM] = fML(I, r, n, x);

    case 6
        % Number of experimental groups.
        I = varargin{1, 1};
        if (prod(size(I)) ~= 1 || (~ isnumeric(I)))
            error('The first parameter has to be a number.');
        end

        % Number of fat pads with no outgrowth per group.
        r = varargin{1, 2};
        if (prod(size(r)) ~= I)
            error('The size of the vector parameters is given by
the first argument.');
        else
            r = reshape(r, 1, I);
        end

        % Total number of fat pads per group.
        n = varargin{1, 3};
        if (prod(size(n)) ~= I)
            error('The size of the vector parameters is given by
the first argument.');
        else
            n = reshape(n, 1, I);
        end

        % Number of cell transplanted per group.
        x = varargin{1, 4};
        if (prod(size(x)) ~= I)
            error('The size of the vector parameters is given by
the first argument.');
        else
            x = reshape(x, 1, I);
        end

        % Confidence interval.
        k = cell2struct(varargin(1, 5), 'k', 1);
        k = k.k;
        if (prod(size(k)) ~= 1 || (~ isnumeric(k)))
            error('The fifth parameter has to be a number.');
        end

        % Estimate of the frequency to measure.
        estimateML = cell2struct(varargin{1, 6}, 'est', 1);
```

```
            estimateML = estimateML.est;
            if (prod(size(estimateML)) ~= 1 || (~
isnumeric(estimateML)))
                  error('The sixth parameter has to be a number.');
            end


      otherwise
            error('fMC requires at least four input arguments');
end

% Total number of fat pads.
N = sum(n, 2);

% Frequency of no outgrowth outcome per group.
p = r ./ n;

estimateMC = zeros(1, 1);

old_est = linspace(estimateML, estimateML, I);

disp('Computing fMC ...');
while (true)
      num = sum((n .* x .* exp(- old_est .* x) .* (2 .* r - n) + r .* r
.* x .* (exp(old_est .* x) - 2)) ./ (n .* ((1 - exp(- old_est .* x)) .^
2)) , 2);
      denom = sum((n .* n .* x .* x .* (exp(- old_est .* x) - exp(- 3
.* old_est .* x)) + n .* r .* x .* x .* (2 .* exp(- 3 .* old_est .* x)
- 2 .* exp(- old_est .* x)) + r .* r .* x .* x .* (exp(old_est .* x) -
4 + 7 .* exp(- old_est .* x) - 4 .* exp(- 2 .* old_est .* x))) ./ (n .*
((1 - exp(- old_est .* x)) .^ 4)), 2);
      correction = num ./ denom;
      new_est = old_est - linspace(correction, correction, I);
      disp(new_est(1));

      if correction < 0.00005 * old_est
            break;
      else
            old_est = new_est;
      end
end

estimateMC = new_est(1);

% Compute 95% confidence intervals for this estimate.
var_estimateMC = 2.0 / sum((n .* n .* x .* x .* (exp(- estimateMC .* x)
- exp(- 3 .* estimateMC .* x)) + n .* r .* x .* x .* (2 .* exp(- 3 .*
estimateMC .* x) - 2 .* exp(- estimateMC .* x)) + r .* r .* x .* x .*
(exp(estimateMC .* x) - 4 + 7 .* exp(- estimateMC .* x) - 4 .* exp(- 2
.* estimateMC .* x))) ./ (n .* ((1 - exp(- estimateMC .* x)) .^ 4)),
2);
ci95 = k * sqrt(var_estimateMC);

disp(cat(2, 'Frequency of positive cells in this population: 1/',
num2str(round(1./estimateMC)), ' (1/', num2str(round(1./(estimateMC +
ci95))), ' - 1/', num2str(round(1./(estimateMC - ci95))), ').'));
```

```
varargout{1, 1} = estimateMC;
varargout{1, 2} = estimateML;

if (exist('estimateWM'))
      varargout{1, 3} = estimateWM;
else
      varargout{1, 3} = [];
end
```

fML.m

```
% [estimateML, estimateWM] = fML(I, r, n, x, estimate)
% Estimates the frequency of "hit-cells" in a limiting dilution
experiment (maximum likelihood method, see [1]).
%
% I = number of cell doses or groups.
% r = number of negative outcomes in each group.
% n = number of repeats in each group.
% x = number of cells used in each group.
% estimate (optional) = an estimate for the value to be computed. If
this parameter is not provided, an estimate
%   is calculated using weighted mean (fWM).
%
% estimateML = estimate of hit cell frequency using maximum likelihood.
% estimateWM = estimate of hit cell frequency using weighted mean or
the user provided estimate.
%
% [1] Taswell C., "Limiting dilution analysis for the determination of
immunocompetent cell frequencies", The Journal of Immunology
%     126(4):1614 (1981).
%
% Rodrigo Fernandez-Gonzalez
% 06/28/2006

function varargout = fMC(varargin)

switch (nargin)
      case 4
            % Number of experimental groups.
            I = varargin{1, 1};
            if (prod(size(I)) ~= 1 || (~ isnumeric(I)))
                  error('The first parameter has to be a number.');
            end

            % Number of fat pads with no outgrowth per group.
            r = varargin{1, 2};
            if (prod(size(r)) ~= I)
                  error('The size of the vector parameters is given by
the first argument.');
            else
                  r = reshape(r, 1, I);
            end

            % Total number of fat pads per group.
            n = varargin{1, 3};
            if (prod(size(n)) ~= I)
```

```
                error('The size of the vector parameters is given by
the first argument.');
            else
                n = reshape(n, 1, I);
            end

            % Number of cell transplanted per group.
            x = varargin{1, 4};
            if (prod(size(x)) ~= I)
                error('The size of the vector parameters is given by
the first argument.');
            else
                x = reshape(x, 1, I);
            end

            % Obtain an estimate of the frequency that is to be
computed.
            estimateWM = fWM(I, r, n, x);

        case 5
            % Number of experimental groups.
            I = varargin{1, 1};
            if (prod(size(I)) ~= 1 || (~ isnumeric(I)))
                error('The first parameter has to be a number.');
            end

            % Number of fat pads with no outgrowth per group.
            r = varargin{1, 2};
            if (prod(size(r)) ~= I)
                error('The size of the vector parameters is given by
the first argument.');
            else
                r = reshape(r, 1, I);
            end

            % Total number of fat pads per group.
            n = varargin{1, 3};
            if (prod(size(n)) ~= I)
                error('The size of the vector parameters is given by
the first argument.');
            else
                n = reshape(n, 1, I);
            end

            % Number of cell transplanted per group.
            x = varargin{1, 4};
            if (prod(size(x)) ~= I)
                error('The size of the vector parameters is given by
the first argument.');
            else
                x = reshape(x, 1, I);
            end

            % Estimate of the frequency to measure.
            estimateWM = varargin{1, 5};
            if (prod(size(estimateWM)) ~= 1 || (~
isnumeric(estimateWM)))
```

```
                error('The fifth parameter has to be a number.');
            end


      otherwise
            error('fML requires at least four input arguments');
end

% Total number of fat pads.
N = sum(n, 2);

% Frequency of no outgrowth outcome per group.
p = r ./ n;

old_est = linspace(estimateWM, estimateWM, I);

disp('Computing fML ...');
while (true)
      num = sum(- r .* x + ((n - r) .* x .* exp(- old_est .* x)) ./ (1
- exp(- old_est .* x)), 2);
      denom = sum((- (n - r) .* x .* x .* exp(- old_est .* x)) ./ ((1 -
exp(- old_est .* x)) .^ 2), 2);
      correction = num ./ denom;
      new_est = old_est - linspace(correction, correction, I);
      disp(new_est(1));

      if correction < 0.00005 * old_est
            break;
      else
            old_est = new_est;
      end
end

estimateML = new_est(1);

varargout{1, 1} = estimateML;
varargout{1, 2} = estimateWM;
```

fWM.m

```
% estimateWM = fWM(I, r, n, x)
% Estimates the frequency of "hit-cells" in a limiting dilution
experiment (weighted mean method, see [1]).
%
% I = number of cell doses or groups.
% r = number of negative outcomes in each group.
% n = number of repeats in each group.
% x = number of cells used in each group.
%
% estimateWM = estimate of hit cell frequency using weighted mean.
%
% [1] Taswell C., "Limiting dilution analysis for the determination of
immunocompetent cell frequencies", The Journal of Immunology
%     126(4):1614 (1981).
%
% Rodrigo Fernandez-Gonzalez
% 06/28/2006
```

```
function varargout = fWM(varargin)

switch (nargin)
      case 4
            % Number of experimental groups.
            I = varargin{1, 1};
            if (prod(size(I)) ~= 1 || (~ isnumeric(I)))
                  error('The first parameter has to be a number.');
            end

            % Number of fat pads with no outgrowth per group.
            r = varargin{1, 2};
            if (prod(size(r)) ~= I)
                  error('The size of the vector parameters is given by
the first argument.');
            else
                  r = reshape(r, 1, I);
            end

            % Total number of fat pads per group.
            n = varargin{1, 3};
            if (prod(size(n)) ~= I)
                  error('The size of the vector parameters is given by
the first argument.');
            else
                  n = reshape(n, 1, I);
            end

            % Number of cell transplanted per group.
            x = varargin{1, 4};
            if (prod(size(x)) ~= I)
                  error('The size of the vector parameters is given by
the first argument.');
            else
                  x = reshape(x, 1, I);
            end

      otherwise
            error('fWM requires four input arguments');
end

disp('Computing fWM ...');

% Total number of fat pads.
N = sum(n, 2);

% Frequency of no outgrowth outcome per group.
p = r ./ n;

% Do not use groups where r = 0 or p = 1
use_indices = find(r ~= 0 & p ~= 1);


% Reassign variables.
I = prod(size(use_indices));
r = r(use_indices);
```

```
n = n(use_indices);
x = x(use_indices);

% Recompute these two.
N = sum(n, 2);
p = r ./ n;

estimateWM = sum(((- log(p)) ./ x) .* (x .* x .* r ./ (1 - p)), 2) /
sum(x .* x .* r ./ (1 - p), 2);

disp(estimateWM);

varargout{1, 1} = estimateWM;
```

References

1. DeOme KB, Faulkin LJJ, Bern HA, Blair PB (1959) Development of mammary tumors from hyperplastic alveolar nodules transplanted into gland-free mammary fat pads of female C3H mice. Cancer Res 19: 515–520
2. Young LJ, Medina D, DeOme KB, Daniel CW (1971) The influence of host and tissue age on life span and growth rate of serially transplanted mouse mammary gland. Exp Gerontol 6:49–56
3. Daniel CW, De Ome KB, Young JT, Blair PB, Faulkin LJ Jr (1968) The in vivo life span of normal and preneoplastic mouse mammary glands: a serial transplantation study. Proc Natl Acad Sci U S A 61:53–60
4. Gould MN, Biel WF, Clifton KH (1977) Morphological and quantitative studies of gland formation from inocula of monodispersed rat mammary cells. Exp Cell Res 107:405–416
5. Kamiya K, Gould MN, Clifton KH (1998) Quantitative studies of ductal versus alveolar differentiation from rat mammary clonogens. Proc Soc Exp Biol Med 219:217–225
6. Kim ND, Oberley TD, Yasukawa-Barnes J, Clifton KH (2000) Stem cell characteristics of transplanted rat mammary clonogens. Exp Cell Res 260:146–159
7. Laidlaw IJ, Clarke RB, Howell A, Owen AW, Potten CS, Anderson E (1995) The proliferation of normal human breast tissue implanted into athymic nude mice is stimulated by estrogen but not progesterone. Endocrinology 136:164–171
8. Clarke RB, Howell A, Anderson E (1997) Estrogen sensitivity of normal human breast tissue in vivo and implanted into athymic nude mice: analysis of the relationship between estrogen-induced proliferation and progesterone receptor expression. Breast Cancer Res Treat 45:121–133
9. Lenth RV (2006) Java Applets for Power and Sample Size (Computer Software] (http://www.stat.uiowa.edu/~rlenth/power)
10. Taswell C (1981) Limiting dilution assays for the determination of immunocompetent cell frequencies I. Data analysis. J Immunol 126: 1614–1619
11. Glantz SA (2005) Primer of biostatistics, McGraw-Hill Medical, New York
12. Fernandez-Gonzalez R (2006) In: Bioengineering, Vol. Ph.D., UC Berkeley/UC San Francisco, Berkeley

Chapter 3

Use of Stem Cell Markers in Dissociated Mammary Populations

Dawne N. Shelton, Rodrigo Fernandez-Gonzalez, Irineu Illa-Bochaca, Carlos Ortiz-de-Solorzano, Mary Helen Barcellos-Hoff, and Bryan E. Welm

Abstract

The regenerative potential of mammary epithelium facilitates assessment of the "stemness" of any epithelial subpopulation in transplantation assays. Thus, mammary tissue can be dissociated into single cells, stained for cell surface markers of interest and classified using fluorescence-activated cell sorting. The selected cells can then be transplanted into epithelium-devoided fat pads from recipient hosts. Recent publications have described markers that enrich for mammary repopulating potential. Here, we describe the materials and methods necessary to sort cells according to these markers. This approach can be used interchangeably with other cell surface markers with slight variation to the protocol.

Key words: Mammary primary culture, CD24, CD49f, Mammary repopulating unit

1. Introduction

The mammary gland contains a hierarchy of differentiated cell types including mammary stem cells, limited-potential progenitors, and differentiated epithelium (1–3). Cell surface markers and fluorescence-activated cell sorting (FACS) have been used to identify and enrich mammary epithelial cell (MEC) populations with distinct regenerative capacity from both normal (stem cells) and tumor (cancer stem cells) tissue (2–4). In addition to enriching for stem cells, cell surface markers can be used to isolate differentiated mammary populations including luminal epithelium, myoepithelium, progenitors, and steroid receptor expressing cells (5–9). FACS sorters can distinguish these cell types by measurement of the relative expression level of cell-surface

Irina M. Conboy et al. (eds.), *Protocols for Adult Stem Cells*, Methods in Molecular Biology, vol. 621,
DOI 10.1007/978-1-60761-063-2_3, © Springer Science+Business Media, LLC 2010

proteins on different epithelial populations. As a result, FACS has become an important technique for studying mammary differentiation and will become used more frequently for the analysis of differentiation phenotypes from genetically engineered mice. While several cell surface markers, such as Sca1, CD24, CD61, CD49f, CD29, and CD133, have been used in different combinations to enrich for distinct mammary cell populations; here, we describe a general FACS method to sort MECs based on expression of CD24 and CD49f (2, 3, 6, 8, 10).

There are a number of important considerations when using FACS to analyze or sort MEC populations. Primary MEC preparations are composed of many nonepithelial cell types such as blood, endothelial, and stromal cells. These contaminating cells can be depleted both during the primary MEC preparation, through differential centrifugation (Fig. 1a–d), and by gating against nonepithelial cell types in FACS (Fig. 1e–h). Additionally, during FACS, cells are gated sequentially to remove both dead and lineage positive cells prior to setting gates for MaCFC, MRU, and Myo, which are MEC populations enriched for progenitors, stem cells, and myoepithelial cells, respectively. This strategy allows for resolution of a clear MRU population (Fig. 1i, J), whereas poor distinction of these populations occurs if cells are only gated based on forward and side scatter (Fig. 1k). Therefore, nonepithelial cells should be depleted prior to analyzing the MRU, MaCFC, and Myo gates during FACS.

Several controls are necessary for FACS analysis in order to determine antibody specificity, gates, photomultiplier tube (PMT) voltage, and compensation. The antibodies used in FACS may interact nonspecifically with FcγIII receptors expressed on several circulating blood cell types and pretreating MECs with antimouse CD16/CD32 antibodies can block this interaction. Nonspecific antibody interactions are also determined by using no-antibody and isotype-antibody controls. The PMT voltage and compensation are established by staining cells individually for each fluorochrome, and these instrument settings can be saved to streamline future setup of the FACS. The following method includes the control samples described above.

2. Materials

1. Hanks balanced salt solution +2% fetal bovine serum (FBS).
2. 15 ml centrifuge tubes.
3. FACS tubes.

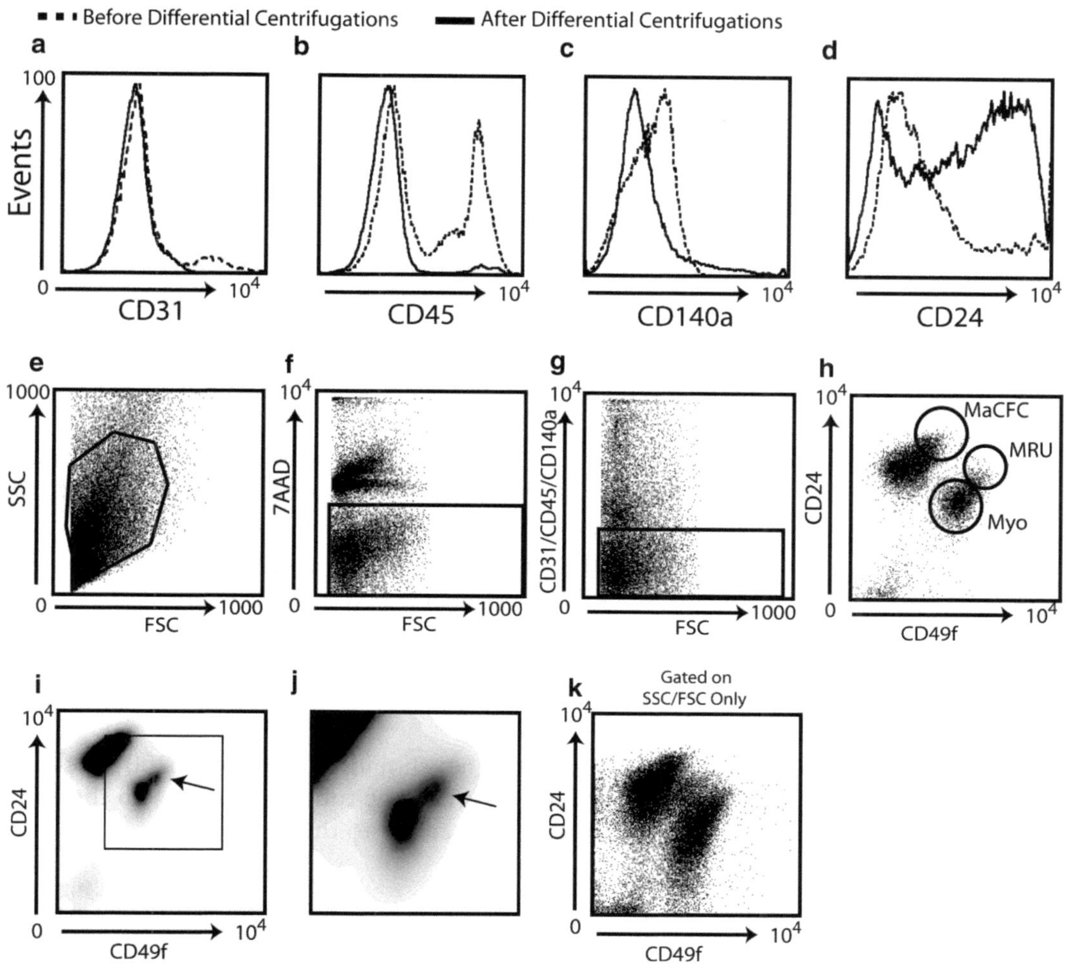

Fig. 1. FACS analysis of CD24/CD49f cell populations in primary MECs. A primary MEC preparation is enriched for different cell populations (**a–d**, *dashed line*) before and (**a–d**, *solid line*) after washes using differential centrifugation. Contaminating cells such as (**a**) endothelial cells, (**b**) blood cells, and (**c**) fibroblasts were depleted, while an increase in the (**d**) $CD24^{+}$-epithelial population was observed after washes, as measured by the cell surface markers CD31, CD45, CD140a, and CD24, respectively. Sequential gating of (**e**) single, (**f**) live, and (**g**) lineage-negative cells results in good resolution of the (**h**) MRU, MaCFC, and Myo gates. (**i** and **j**) A density plot of sequentially gated MECs allows distinction of the MRU population (*arrow*). Panel **j** shows a close up view of the boxed area in panel **i**. MECs that are only gated on side and forward scatter causes poor resolution of the MRU, MaCFC, and Myo cell populations (compare **h** and **k**)

4. BD FACScan with 488 nm-argon and 637 nm-red-diode lasers with the following filter/detector setup: FL1 = 530/30 nm, FL2 = 585/42 nm, FL3 = 668 nm long pass, FL4 = 666/27 nm.
5. FlowJo Software (Tree Star, Inc).

3. Methods

3.1. Staining Primary MECs for FACS Analysis

1. Perform single cell isolations as described in Chapter 2, Subheading 3.1. and keep the cells on ice and in the dark during all subsequent steps. The HBSS +2%FBS solution should also be kept on ice.
2. Resuspend the cells so that there are 5×10^6 cells per 1 ml HBSS +2%FBS.
3. Block Fc receptors:
   (a) Add 0.5 μg of antimouse CD16/CD32 for each 1×10^6 cells.
   (b) Incubate 5 min on ice. There is no need to wash the cells after this blocking step, proceed directly to step 4 (see Note 1).
4. Aliquot 200 μl (1×10^6 cells) of the cell suspension into the eight tubes listed below. We prefer using 15 ml conical tubes so that we can perform the centrifugation steps with a swinging bucket rotor. The cell number can be reduced in subsequent experiments once the sorting gates and approximate PMT voltages are established.
   (a) The following tubes should be setup with 200 μl of cells (1×10^6 cells) in each. The tubes are listed with the antibodies that will be added to them in step 5.
      i. No antibody.
      ii. Rat IgG2a-FITC and Rat IgG2b-FITC (isotype controls).
      iii. CD49f-FITC (FL1 setup).
      iv. CD24-PE (FL2 setup).
      v. 7AAD (FL3 setup).
      vi. CD24-biotin-streptavidin-APC (FL4 setup).
      vii. Streptavidin-APC (specificity control).
      viii. CD49f-FITC/CD24-biotin-streptavidin-APC/CD31-PE/CD45-PE/CD140a-PE /7AAD (experimental sample).
5. Add the appropriate concentration of antibodies to the proper tube (see Table 1).
6. Incubate the cells on ice for 15 min. (see Note 2).
7. After the incubation, add 1 ml HBSS +2% FBS to each tube and centrifuge at approximately 1,000 × *g* for 2 min. If using Eppendorf tubes with a fixed angle rotor, the cells may not pellet at the bottom but may be attached to the wall of the tube. Be sure to rinse the cells off the tube wall if needed.

Table 1
Antibodies and staining reagents for FACS analysis

Name	Source	Catalog number	Concentration
Rat anti-CD16/CD32	BD Bioscience	553141	0.5 μg/10^6 cells
Rat IgG2a-FITC	BD Bioscience	553929	1:100 vol/vol
Rat IgG2b-FITC	BD Bioscience	553988	1:100 vol/vol
Rat anti-CD49f-FITC	BD Bioscience	555735	1:50 vol/vol
Rat anti-CD24-PE	BD Bioscience	553262	1:100 vol/vol
Rat anti-CD24-biotin	BD Bioscience	553260	1:100 vol/vol
Streptavidin-APC	BD Bioscience	554067	1:200 vol/vol
Rat anti-CD31-PE	BD Bioscience	553373	1:100 vol/vol
Rat anti-CD45-PE	BD Bioscience	553081	1:100 vol/vol
Rat anti-CD140a-PE	eBioscience	121401	1:100 vol/vol
7AAD	BD Bioscience	559925	0.25 μg/10^6 cells

8. Add 500 μl of HBSS +2% FBS to the cells and centrifuge the cells again at 1000 × g for 2 min.
9. Resuspend the cells stained with CD24-biotin in 200 μl of HBSS +2% FBS and add streptavidin-APC (1:200 vol/vol). Incubate on ice for 15 min. The cells in the remaining tubes can be resuspended in 400 μl HBSS +2%FBS.
10. Repeat wash steps 7 and 8 for the CD24 stained cells and add 400 μl HBSS +2%FBS to the final pellet.
11. Add 7AAD to the appropriate tubes (v and viii) at 0.25 μg/1 × 10^6 cells.
12. Let the cells incubate on ice for 15 min and then bring them to the FACS analyzer.

3.2. FACS Machine Setup

1. Use tubes i–vii to determine antibody specificity and set gates, PMT voltage, and compensation. For analysis of MRU, MaCFC, and Myo populations in tube viii, use the sequential gating listed below and shown in Fig. 1 e–h.
   (a) SSC vs. FSC – *gate single cells and discriminate against debris.*
   (b) FL3 (7AAD) vs. FSC – *gate live cells (7AAD negative).*

(c) FL2 (CD31, CD45, CD140a) vs. FSC – *gate lineage negative cells.*

(d) FL4 (CD24-APC) vs. FL1 (CD49f-FITC) – *gate MRU, Myo, and MaCFC.*

4. Notes

1. If you plan on using a Rat IgG primary antibody with a conjugated secondary antibody, the Fc block will cross-react with anti-rat-IgG secondary antibodies.
2. For CD24 staining (tubes vi and viii), only add the CD24-biotin antibody at this stage.

Acknowledgments

This work was supported by a predoctoral fellowship to RFG from the Department of Defense Breast Cancer Research Program (DAMD 17-03-1-0594), grants from the same institution to COS (DAMD 17-00-1-0227 and DAMD 17-00-1-0306), a grant to BEW from the National Cancer Institute (CA 8424306), and a grant to MHBH funded by the National Institute of Environmental Health Sciences and the National Cancer Institute (U01ES012801).

References

1. Smith GH (1996) Experimental mammary epithelial morphogenesis in an in vivo model: evidence for distinct cellular progenitors of the ductal and lobular phenotype. Breast Cancer Res Treat 39:21–31
2. Shackleton M, Vaillant F, Simpson KJ, Stingl J, Smyth GK, Asselin-Labat ML, Wu L, Lindeman GJ, Visvader JE (2006) Generation of a functional mammary gland from a single stem cell. Nature 439:84–88
3. Stingl J, Eirew P, Ricketson I, Shackleton M, Vaillant F, Choi D, Li HYI, Eaves CJ (2006) Purification and unique properties of mammary epithelial stem cells. Nature 439:993–997
4. Al-Hajj M, Wicha MS, Benito-Hernandez A, Morrison SJ, Clarke MF (2003) Prospective identification of tumorigenic breast cancer cells. Proc Natl Acad Sci U S A 100:3983–3988
5. Asselin-Labat ML, Shackleton M, Stingl J, Vaillant F, Forrest NC, Eaves CJ, Visvader JE, Lindeman GJ (2006) Steroid hormone receptor status of mouse mammary stem cells. J Natl Cancer Inst 98:1011–1014
6. Asselin-Labat ML, Sutherland KD, Barker H, Thomas R, Shackleton M, Forrest NC, Hartley L, Robb L, Grosveld FG, van der Wees J, Lindeman GJ, Visvader JE (2007) Gata-3 is an essential regulator of mammary-gland morphogenesis and luminal-cell differentiation. Nat Cell Biol 9:201–209
7. Sleeman KE, Kendrick H, Ashworth A, Isacke CM, Smalley MJ (2006) CD24 staining of mouse mammary gland cells defines luminal epithelial, myoepithelial/basal and non-epithelial cells. Breast Cancer Res 8:R7
8. Sleeman KE, Kendrick H, Robertson D, Isacke CM, Ashworth A, Smalley MJ (2007) Dissociation of estrogen receptor expression and in vivo stem cell activity in the mammary gland. J Cell Biol 176:19–26

9. Stingl J, Eaves CJ, Zandieh I, Emerman JT (2001) Characterization of bipotent mammary epithelial progenitor cells in normal adult human breast tissue. Breast Cancer Res Treat 67:93–109

10. Welm BE, Tepera SB, Venezia T, Graubert TA, Rosen JM, Goodell MA (2002) Sca-1(pos) cells in the mouse mammary gland represent an enriched progenitor cell population. Dev Biol 245:42–56

Chapter 4

Isolation of Adult Hippocampal Neural Progenitors

Joseph Peltier, Brandi K. Ormerod, and David V. Schaffer

Abstract

Adult neurogenesis, or the creation of new neurons in adult organisms, is an exciting recent area of neurological research. The subgranular zone of the adult hippocampus is one area where hippocampal neural progenitors generate new neurons that functionally integrate into existing neuronal circuitry. Given the role that the hippocampus plays in learning and memory consolidation and its vulnerability to neurological diseases and conditions, such as Alzheimer's disease, understanding the mechanisms controlling the self-renewal and differentiation of neural progenitor cells is a critical first step in developing novel disease treatments. In this and subsequent chapters, we describe many of the *in vivo* and *in vitro* techniques necessary to study hippocampal progenitors in the adult rat. Specifically, this chapter details isolation of progenitors from the adult rat for the establishment of *in vitro* culture.

Key words: Neural, Hippocampus, Progenitors, Rat, *In vitro*, Adult, Isolation

1. Introduction

The discoveries that new neurons are generated continuously in some adult brain regions (adult neurogenesis) (1), and that the progenitor cells that generate these new neurons can be harvested and grown in culture (2, 3), have created exciting new possibilities for the treatment of many neurological diseases and injuries. *In vitro*, adult hippocampal neural progenitor cells (AHNPCs) can generate nearly all major cell types within the mammalian brain, including neurons, astrocytes, and oligodendrocytes, which makes them promising candidates for the treatment of neurological injuries and diseases. Additionally, neurogenesis may play roles in learning and memory (4–6), the effects of exercise on learning (6), stress and depression (7, 8), response to injury (9, 10), and aging (11). Therefore, understanding AHNPC behavior is important for developing the strategies needed to efficiently generate neurons,

Irina M. Conboy et al. (eds.), *Protocols for Adult Stem Cells*, Methods in Molecular Biology, vol. 621,
DOI 10.1007/978-1-60761-063-2_4, © Springer Science+Business Media, LLC 2010

astrocytes, oligodendrocytes, or more AHNPCs for treatment of neurological disease or injury.

Here, we describe a protocol for the isolation of hippocampal progenitors from the adult rat brain, which is accomplished via a two-step cell fractionation process. Dissociated hippocampi from several rats are pooled and fractionated in a 35% Percoll solution. This first centrifugation separates progenitors and red blood cells (pellet) from other differentiated neural cell types (supernatant). The second fractionation step occurs in 65% Percoll and separates the red blood cells (pellet) from progenitors (supernatant). Once isolated, the progenitors are plated on poly-L-ornithine/laminin-coated culture dishes and can be repeatedly passaged. These cells have stable karyotype for up to 35 population doublings in culture (2) and can readily be cryo-preserved for long-term storage. See Chapter 5 for passaging and cryo-preservation protocols. Protocols for analysis of proliferation, immunostaining, and quantitative RT-PCR are also found in Chapter 5. Chapter 6 details *in vivo* engraftment and analysis, and protocols for viral production and transduction of AHNPCs are found in Chapter 7.

2. Materials

1. 3–6 female Fischer 344 rats (6-week-old; Charles River Laboratories, Wilmington, MA).
2. Medium Animal Guillotine (DCAP, 2.5″ opening, World Precision Instruments, Sarasota, FL).
3. Sterile scalpel blades (No. 10).
4. Dulbecco's Phosphate Buffered Saline (DPBS) w/o calcium/magnesium (Invitrogen, Bethesda, MD) + 1 g/L glucose. Sterilize through a 0.22 μm filter and chill on ice before use.
5. Papain, suspension (Worthington Biochemical, Lakewood, NJ). Store 4°C.
6. Deoxyribonuclease I, lyophilized (Worthington Biochemical). Store 4°C.
7. Dispase II (Roche Applied Science, Indianapolis, IN). Store 4°C.
8. Dulbecco's Modified Eagle Medium/Nutrient Mix F-12 (DMEM/F-12) with HEPES and L-Glutamine (Invitrogen).
9. Enzyme Cocktail: Dissolve 0.47 ml papain (2.5 U/ml), 19.98 mg deoxyribonuclease I (250 U/ml), and 400 mg Dispase II (1 U/ml) into 200 ml DMEM/F-12. Sterilize through a 0.22 μm filter, and freeze at −20°C in 50 ml aliquots. Warm to 37°C for approximately 30 min before use to fully activate the papain (see Note 1). Avoid freeze/thaw.

10. Percoll™ (Amersham, Piscataway, NJ). Store at 4°C.
11. 10× DPBS w/o calcium/magnesium (Invitrogen).
12. Percoll™ solution (100% Percoll™): Dilute 5 ml 10× DPBS into 45 ml Percoll™. Use immediately.
13. Fetal bovine serum (FBS, Invitrogen).
14. Harvest media: DMEM/F-12 containing 10% (v/v) FBS.
15. Trypan blue stain (Invitrogen).
16. Hemacytometer.
17. 3.5 cm poly-L-ornithine/laminin coated culture plate (see Subheading 2.2 for preparation).
18. DMEM/F-12 supplemented with 1% (v/v) N-2 Supplement (Invitrogen). Store supplemented medium at 4°C for up to 1 month.
19. Basic fibroblast growth factor (FGF-2, Peprotech, Rocky Hill, NJ).
20. FGF-2 stock solution: dissolve in DMEM/F-12 + N-2 at 0.1 mg/ml, aliquot, and store at –20°C. Once thawed, aliquot can be stored at 4°C for up to 1 week.

2.1. Anesthesia

1. Ketamine Hydrochloride, 100 mg/ml (Phoenix Scientific, St. Joseph, MO).
2. Xylazine Hydrochloride, 20 mg/ml (Ben Venue Laboratories, Bedford, OH).
3. 9% NaCl (w/v) in distilled/de-ionized water. Sterilize by 0.22 μm filtration.
4. Anesthetic cocktail: For each 150 g rat, sterilely mix 135 μl ketamine, 75 μl xylazine, 50 μl 9% NaCl, and 240 μl distilled/de-ionized water.

2.2. Laminin Coated Culture Plates

2.2.1. Reagents

1. Sterile tissue culture water.
2. Sterile phosphate buffered saline (PBS), pH 7.4 (Invitrogen).
3. Poly-L-ornithine hydrobromide (Sigma, St. Louis, MO) dissolved in tissue culture water at 10 mg/ml. Sterilize by 0.22 μm syringe filtration, aliquot, and store at –20°C. Avoid repeated freeze/thaw.
4. Natural mouse laminin (Invitrogen). Store at –80°C. Thaw slowly at 4°C before use. Avoid repeated freeze/thaw and vortexing.
5. Plastic wrap.

2.2.2. Preparation of Plates

1. Dilute poly-L-ornithine in tissue culture water to a final concentration of 10 μg/ml and add to tissue culture plates such that plates are well covered (~2 ml for 3.5-cm plate). Incubate overnight at room temperature. To maintain sterility,

leave the plates in a biological safety cabinet with blower off and sash closed. Leave UV light OFF.

2. Rinse plates twice with approximately the same amount of sterile water as used for poly-L-ornithine incubation. Do not allow plates to dry out.
3. Dilute laminin in sterile PBS to a final concentration of 5 μg/ml and add to plates at the same volume as was used for poly-L-ornithine.
4. Incubate at 37°C overnight.
5. Wrap in plastic wrap and store at –20°C for up to 6 months. Freeze plates on a level surface to avoid pooling and dry spots.

3. Methods

1. Anesthetize 3–6-week-old female Fischer 344 rats using ketamine/xylazine anesthetic cocktail (500 μL of anesthesia per 150 g of body weight) via intraperitoneal injection (see Note 2).
2. Wait several minutes to ensure that the rat is anesthetized. Check reflexes by pulling the hind limb laterally and pinching the hind paw while the animal is on its stomach. If the rat is not fully under it will retract its hind paw.
3. Decapitate the animals using the rodent guillotine, and as aseptically as possible, carefully remove the whole brain from the skull. Place brains in 50 mL centrifuge tubes containing 25 mL ice-cold DPBS w/o calcium/magnesium + 1 g/L glucose (see Note 3).
4. In a biological safety cabinet with autoclaved scalpels remove the hippocampi as aseptically as possible. Bisect each brain longitudinally and separate each hippocampal lobe from the cortical white matter using the natural separation along the alveus hippocampus. Remove as much white matter as possible (see Note 4).
5. In a sterile, dry tissue culture plate, finely mince the hippocampi using two sterile scalpel blades. Properly minced tissue should be easily aspirated into a sterile 10 ml pipette.
6. Add 3 ml enzyme cocktail per hippocampal lobe to 50 ml centrifuge tubes and warm in 37°C water bath for 30 min.
7. Transfer minced tissue into the warmed enzyme cocktail with a sterile 10 ml pipette. Incubate in 37°C water bath for 45 min with occasional gentle mixing. Tissue chunks will still be present, but the solution should be slightly turbid.

8. Centrifuge at 500 g for 5 min. Aspirate and discard enzyme cocktail supernatant.
9. Resuspend cells in 10 ml harvest medium and transfer to sterile 15 ml tube.
10. Centrifuge at 500 g for 5 min. Aspirate and discard supernatant.
11. Repeat Steps 9 and 10.
12. Resuspend cells with 10mL 35% Percoll solution in harvest medium. Thoroughly mix Percoll solution with cells and medium. Do not layer.
13. Centrifuge at 800 g for 10 min without brake. Progenitors are now in the pellet with red blood cells. Aspirate and discard supernatant.
14. Resuspend pellet with 10mL 65% Percoll solution in harvest medium.
15. Centrifuge at 800 g for 10 min without brake. Progenitors are now in suspension while the pellet primarily contains red blood cells.
16. Pipette supernatant into sterile 50 ml centrifuge tubes, noting the volume pipetted into each tube. Dilute the cell suspension with 4× volume of harvest media to decrease the buoyancy of the progenitors.
17. Centrifuge at 800 g for 10 min. Aspirate and discard supernatant.
18. Resuspend all pellets in a total of 10 ml harvest medium. Transfer to 15 ml centrifuge tube. Centrifuge at 800 g for 5 min. Carefully aspirate and discard supernatant.
19. Resuspend in 10 ml harvest medium.
20. Centrifuge at 800 g for 5 min. Aspirate and discard supernatant.
21. Repeat Steps 19 and 20. These pellets will be quite small and loose, so use care when aspirating supernatant to avoid disturbing the pellet.
22. Resuspend pellet in 2 ml DMEM/F-12 + N-2 and quantify cell concentration by hemacytometer using 1:1 DPBS/trypan blue solution to assess viability.
23. Thaw 3.5 cm poly-L-ornithine/laminin coated culture dish and aspirate PBS. Immediately plate well-mixed cells at approximately 10^5 live cells per cm^2 (2 ml total volume). Do not allow the plate to dry out. Gently rock plate to evenly disperse cells. Do not swirl.
24. Supplement the culture with 20 ng/ml FGF-2 (0.2 μl of FGF-2 stock per ml of culture). Gently rock the plate to mix.
25. Incubate cells at 37°C in 5% CO_2 atmosphere.

26. Replace 50% of the medium with fresh, warm DMEM/F-12 + N-2 + 40 ng/ml FGF-2 every other day.
27. Cells should be ~80% confluent and ready to passage within 3–10 days. See Chapter 5 Subheading 3.1 for passaging protocol.

4. Notes

1. Papain is a cysteine-dependent protease. DMEM/F-12 contains the cysteine required for full papain activation.
2. Alternatively, CO_2 asphyxiation may be used for anesthetization followed by cervical dislocation.
3. If contamination is a concern, then 7.5 μg/ml of Fungizone® (Invitrogen) + 3% (v/v) penicillin–streptomycin (10,000 U/ml, Invitrogen) may be added to the DPBS.
4. A dissecting microscope may be used for improved visualization of the appropriate brain structures during dissection.

Acknowledgments

We acknowledge Theo Palmer for helpful technical comments and Matthew Robertson for critical reading of the manuscript. This work was funded by an NSF Graduate Research Fellowship (to JP), a California Institute for Regenerative Medicine Training Grant (T1-00007), and NSF BES-0629202.

References

1. Altman J (1962) Are new neurons formed in the brains of adult mammals? Science 135: 1127–1128
2. Palmer TD, Takahashi J, Gage FH (1997) The adult rat hippocampus contains primordial neural stem cells. Mol Cell Neurosci 8:389–404
3. Ray J, Peterson DA, Schinstine M, Gage FH (1993) Proliferation, differentiation, and long-term culture of primary hippocampal neurons. Proc Natl Acad Sci USA 90: 3602–3606
4. Kempermann G, Gage FH (1999) Experience-dependent regulation of adult hippocampal neurogenesis: effects of long-term stimulation and stimulus withdrawal. Hippocampus 9:321–332
5. Kempermann G, Kuhn HG, Gage FH (1997) More hippocampal neurons in adult mice living in an enriched environment. Nature 386:493–495
6. van Praag H, Christie BR, Sejnowski TJ, Gage FH (1999) Running enhances neurogenesis, learning, and long-term potentiation in mice. Proc Natl Acad Sci USA 96: 13427–13431
7. Gould E, Reeves AJ, Fallah M, Tanapat P, Gross CG, Fuchs E (1999) Hippocampal neurogenesis in adult Old World primates. Proc Natl Acad Sci USA 96:5263–5267
8. Santarelli L, Saxe M, Gross C, Surget A, Battaglia F, Dulawa S, Weisstaub N, Lee J, Duman R, Arancio O, Belzung C, Hen R (2003) Requirement of hippocampal

neurogenesis for the behavioral effects of antidepressants. Science 301:805–809

9. Liu J, Solway K, Messing RO, Sharp FR (1998) Increased neurogenesis in the dentate gyrus after transient global ischemia in gerbils. J Neurosci 18:7768–7778
10. Takagi Y, Nozaki K, Takahashi J, Yodoi J, Ishikawa M, Hashimoto N (1999) Proliferation of neuronal precursor cells in the dentate gyrus is accelerated after transient forebrain ischemia in mice. Brain Res 831: 283–287
11. Gould E, Beylin A, Tanapat P, Reeves A, Shors TJ (1999) Learning enhances adult neurogenesis in the hippocampal formation. Nat Neurosci 2:260–265

Chapter 5

In Vitro Culture and Analysis of Adult Hippocampal Neural Progenitors

Joseph Peltier, Smita Agrawal, Matthew J. Robertson, and David V. Schaffer

Abstract

Adult hippocampal neural progenitor cell (AHNPC) culture is a useful technique for gaining insight into adult neurogenesis, studying disease, and high throughput drug screening. The ability of AHNPCs to proliferate and differentiate into the three cell lineages of the adult brain in cell culture provides the researcher a powerful platform to study the extracellular and intracellular regulatory mechanisms in a well-controlled environment. In this chapter, we describe some of the *in vitro* techniques necessary to study hippocampal progenitors in the adult rat. This chapter details routine culture techniques for passaging and differentiating hippocampal progenitors. We also describe techniques for analyzing the culture state, such as proliferation and expression of cell fate markers by quantitative RT-PCR and immunofluorescence.

Key words: Neural, Hippocampus, Progenitors, Rat, *In vitro*, Adult, Quantitative RT-PCR, Immunostaining, Differentiation, Proliferation, Neurons, Astrocytes

1. Introduction

The ability to culture and analyze hippocampal neural progenitor cells *in vitro* is crucial to researching their function and behavior in the adult brain. *In vitro*, adult hippocampal neural progenitor cells (AHNPCs) can undergo self-renewal as well as generate nearly all major cell types in the mammalian brain, including neurons, astrocytes, and oligodendrocytes (1). Therefore, culturing these cells and analyzing them in a culture dish provides a facile, initial means of testing hypotheses before conducting *in vivo* experiments that are often resource-intensive and time-consuming.

Here, we describe the protocols necessary for culture, differentiation, and analysis of AHNPCs *in vitro* (see Chapter 4

Irina M. Conboy et al. (eds.), *Protocols for Adult Stem Cells*, Methods in Molecular Biology, vol. 621,
DOI 10.1007/978-1-60761-063-2_5, © Springer Science+Business Media, LLC 2010

for isolation of primary culture from the adult rat). Once derived, these cells have stable karyotype for up to 35 population doublings (2) and can readily be cryo-preserved for long-term storage. We also describe typical techniques for the *in vitro* analysis of cell behavior. These include proliferation and expression of cell fate markers as analyzed by quantitative RT-PCR and immunofluorescence.

We have used these techniques to investigate the inter- and intra-cellular signals responsible for AHNPC proliferation, self-renewal, and differentiation (3, 4). When designing an experimental plan involving these assays, it is especially important to be familiar with the strengths and weaknesses of each technique. Often, more than one of these techniques is required to convincingly demonstrate the existence of a phenomenon. A proliferation assay is not a measure of multipotency. Rather, it merely provides a measure of cell number and therefore growth rate. Other assays, such as quantitative RT-PCR and/or immunofluorescence for a multipotent maker, must accompany proliferation results to demonstrate multipotency. Additionally, quantitative RT-PCR measures a population-averaged concentration of mRNA in an entire culture. If cell-to-cell variability within a culture is important, then we recommend analyzing the culture via immunostaining to gain single cell data. Another key difference between quantitative RT-PCR and immunofluorescence is that the former measures mRNA concentration while the latter measures protein concentration. The two concentrations may not necessarily follow the same trend.

Many of the culture techniques outlined here are necessary for the completion of protocols detailed in other chapters describing the isolation (Chapter 4), *in vivo* engraftment and analysis (Chapter 6), and viral transduction (Chapter 7) of AHNPCs. Therefore, we recommend that the reader become comfortable with the routine culture described here before proceeding to the other chapters.

2. Materials

2.1. Culture, Cryo-Preservation, and Differentiation of Neural Progenitors

2.1.1. Laminin-Coated Culture Plates

2.1.1.1. Reagents

1. Sterile tissue culture water.
2. Sterile phosphate buffered saline (PBS), pH 7.4 (Invitrogen, Bethesda, MD).
3. Poly-L-ornithine hydrobromide (Sigma, St. Louis, MO) dissolved in tissue culture water at 10 mg/ml. Sterilize by 0.22 µm syringe filtration, aliquot, and store at –20°C. Avoid repeated freeze/thaw.

4. Natural mouse laminin (Invitrogen). Store at −80°C. Thaw slowly at 4°C before use. Avoid repeated freeze/thaw and vortexing.
5. Plastic wrap.

2.1.1.2. Preparation of Plates

6. Dilute poly-L-ornithine in tissue culture water to a final concentration of 10 μg/ml and add to tissue culture plates such that plates are well covered (~7 ml for 10 cm plate). Incubate overnight at room temperature. To maintain sterility, leave the plates in a biological safety cabinet with blower off and sash closed. Leave UV light OFF.
7. Rinse plates twice with approximately the same amount of sterile water as used for poly-L-ornithine incubation. Do not allow plates to dry out.
8. Dilute laminin in sterile PBS to a final concentration of 5 μg/ml and add to plates at the same volume as was used for poly-L-ornithine.
9. Incubate at 37°C overnight.
10. Wrap in plastic wrap and store at −20°C for up to 6 months. Freeze plates on a level surface to avoid pooling and dry spots.

2.1.2. Cell Culture, Proliferation, and Differentiation Reagents

1. Dulbecco's Modified Eagle Medium/Nutrient Mix F-12 (DMEM/F-12) with HEPES and L-Glutamine (Invitrogen) supplemented with 1% (v/v) N-2 Supplement (Invitrogen). Store supplemented medium at 4°C for up to 1 month.
2. Accutase (Innovative Cell Technologies, San Diego, CA).
3. Basic fibroblast growth factor (FGF-2, Peprotech, Rocky Hill, NJ). FGF-2 stock solution: dissolve in DMEM/F-12 + N-2 at 0.1 mg/ml, aliquot, and store at −20°C. Once thawed, aliquot can be stored at 4°C for up to 1 week.
4. Poly-L-ornithine/laminin coated culture plates (see Subheading 2.1.1).
5. Sterile cotton-plugged glass Pasteur pipettes.
6. 2× freezing medium: 20% (v/v) DMSO dissolved in DMEM/F-12 + N-2. Use immediately.
7. Fetal bovine serum (FBS).
8. Retinoic acid (RA), all-trans (Biomol, Plymouth Meeting, PA). Reconstitute to 1 mM in DMSO. Aliquot and store at −80°C protected from light. Avoid freeze/thaw. Reconstituted RA is stable for up to 3 months at −80°C. Use care when handling and disposing, as RA is a potent toxin and teratogen.
9. Forskolin (Biomol). Reconstitute to 5 mM in DMSO. Aliquot and store at −20°C protected from light. Avoid freeze/thaw. Reconstituted forskolin is stable for up to 2 months at −20°C.

10. Mixed population differentiation medium: 1 μM RA (1 μl RA stock per ml) + 1% (v/v) FBS (10 μl/ml) in DMEM/F-12 + N-2. Use immediately.
11. Neuronal differentiation medium: 1 μM RA (1 μl RA stock per ml) + 5 μM forskolin (1 μl forskolin stock per ml) in DMEM/F-12 + N-2. Use immediately.
12. Optional: Hemacytometer.

2.2. Analyzing Culture State

2.2.1. Proliferation Assay

1. All reagents in Subheading 2.1.2 items 1–5; however, DMEM/F-12 + N-2 should be without phenol red to avoid interference with the colorimetric WST-1 Reagent.
2. Tissue culture plate (96-well) coated with poly-L-ornithine/laminin (see Subheading 2.1.1).
3. Hemacytometer.
4. Cell Proliferation Reagent WST-1 (Roche, Indianapolis, IN). Upon receipt, thoroughly thaw at 37°C and swirl/pipette to ensure all aggregates dissolve. Then aliquot and store at −20°C. Protect from light. Aliquots can be freeze/thawed if necessary. Before use, warm again to 37°C to dissolve aggregates.
5. Multichannel pipettes appropriate for 10–100 μl volumes.
6. Microplate reader (96-well).

2.2.2. Immunostaining

1. Poly-L-ornithine/laminin-coated (see Subheading 2.1.1) CultureSlides (BD Falcon, Bedford, MA). We recommend 8-well CultureSlides, and the volumes used here are for the 8-well format; however, other formats may be used with properly adjusted volumes.
2. PBS, pH 7.4 (Invitrogen).
3. 10× PBS, pH 7.4 (Invitrogen).
4. 4% (w/v) paraformaldehyde. While stirring, carefully heat 400 ml water on a hot plate in a fume hood to 60°C. Do not heat above 60°C, because aldehyde will form and cause autofluorescence. While maintaining temperature, add 20 g paraformaldehyde and stir. As necessary, add NaOH pellets one at a time until paraformaldehyde dissolves. Generally, 1–2 NaOH pellets over the course of 10–15 min will fully dissolve the paraformaldehyde. There will still be small particulates in suspension. Cool to room temperature. Add 50 ml 10× PBS and pH to 7.4 with HCl. Bring total volume to 500 ml with water and sterile filter to remove particulates. Paraformaldehyde must be disposed as chemical waste.
5. Triton X-100 (EMD Biosciences, San Diego, CA).
6. Donkey serum (Sigma), or other appropriate normal serum for blocking (see Note 1). Upon receipt, thaw at

Table 1
Recommended primary antibodies, dilutions, and storage conditions for immunostaining

Name	Species	Supplier	Dilution	Storage
β-Tubulin III	Mouse	Sigma	1:250	Aliquot −20°C
MAP2a + 2b	Mouse	Sigma	1:250	Aliquot −20°C
GFAP	Rabbit	Abcam (Cambridge, MA)	1:500	4°C
O4 (see Note 16)	Mouse	Millipore (Billerica, MA)	1:250	4°C
Nestin	Mouse	BD Pharmingen (San Jose, CA)	1:1,000	4°C

room temperature, aliquot, and store at −20°C. Avoid freeze/thaw.

7. Primary antibodies. See Table 1 for recommended antibodies, dilutions, and storage conditions.
8. Cy3 and Cy5 conjugated donkey anti-rabbit, and antibodies (Jackson ImmunoResearch, West Grove, PA). Upon receipt, reconstitute in 1:1 glycerol:water according to package insert (volume varies by lot), aliquot, and store at −20°C protected from light. Avoid freeze/thaw.
9. TO-PRO®-3 nuclear stain (Invitrogen). Upon receipt, aliquot and store at −20°C. Protect from light and avoid freeze/thaw.
10. Alexa Fluor® 488 conjugated donkey anti-mouse and anti-rabbit antibodies (Invitrogen). Upon receipt, aliquot and store at −20°C protected from light. Avoid freeze/thaw.
11. Microscope cover glasses (24 × 50 mm, Fisher, Pittsburgh, PA).
12. Clear nail varnish.
13. ProLong® Antifade Kit (Invitrogen).

2.2.3. Quantitative RT-PCR

2.2.3.1. RNA Isolation

1. RNase Away (Molecular BioProducts, San Diego, CA).
2. RNase-free, presterilized pipette tips (e.g., ART® sterile, self-sealing, barrier pipette tips, Sigma).
3. RNase-free water (Invitrogen).
4. RNase-free micro-centrifuge tubes (Fisher).
5. TRIzol® Reagent (Invitrogen). Causes burns. Always use gloves when handling, and dispose of residual with chemical waste.
6. Cell Lifter (Corning, Corning, NY).
7. Chloroform (ACS Grade).
8. DNase I, RNase-free, 10 U/μl (Roche, Indianapolis, IN).

9. Glycogen for molecular biology, 20 mg/ml (Roche).
10. Isopropanol (ACS Grade).
11. 75% ethanol, prepared with RNase-free water.

2.2.3.2. Reverse Transcription

12. ThermoScript™ RT-PCR System (Invitrogen) containing:
    ThermoScript™ reverse transcriptase
    5× cDNA synthesis buffer
    0.1 M dithiothreitol (DTT)
    10 mM dNTP mix
    RNase OUT™
    Random hexamers primer
    DEPC-treated water
    RNase H
13. RNase Away (Molecular Biology Products).
14. RNase-free, presterilized pipette tips (e.g., ART® sterile, self-sealing, barrier pipette tips, Sigma).

2.2.3.3. QPCR

15. Distilled/de-ionized water (Invitrogen).
16. Antibody/Taq polymerase. Mix Taq DNA polymerase (Invitrogen) and Anti-Taq JumpStart™ Antibody (Sigma) in a 1:1 ratio. Incubate at room temperature for 10 min, aliquot, and store at −20°C.
17. Taq DNA polymerase 10× PCR Buffer (Invitrogen). Aliquot and store at −20°C.
18. 25 mM $MgCl_2$ (Thermo Scientific, Waltham, MA). Aliquot and store at −20°C.
19. 10 mM dNTP mix. Formulated from 100 mM dNTP Set (Invitrogen). Mix all four dNTPs and dilute with distilled/de-ionized water such that the final concentration of each dNTP is 10 mM. Store at −20°C.
20. 2× PCR supermix. For 1 mL, mix 200 μl 10× PCR Buffer, 240 μl 25 mM $MgCl_2$, 40 μl 10 mM dNTP mix, and 520 μl distilled/de-ionized water. Can be kept at −20°C for long-term storage.
21. Real-time PCR detection system (e.g., iQ5, Bio-Rad Laboratories, Hercules, CA).
22. Thermowell® 96-well PCR Plate (Corning).
23. Microseal B Adhesive Seals (Bio-Rad).
24. Primer and probe sets specific for the gene(s) of interest (Biosearch Technologies, Novato, CA). All primer and probe stocks should be diluted to 10 μM in distilled/de-ionized water. Table 2 lists primer and probe sets for common differentiation markers.

Table 2
Primers, probes, and amplicons for quantitative PCR

Name	Sequence	T_M (°C)
β-Tubulin III		
Sense primer	5′-GCATGGATGAGATGGAGTTCACC-3′	65.2
Antisense primer	5′-CGACTCCTCGTCGTCATCTTCATAC-3′	65.4
Probe	5′-FAM490-TGAACGACCTGGTGTCTGAG-BHQ-3′	59.9
Amplicon	GCATGGATGAGATGGAGTTCACCGAGGCCGAGAGCAACATG AACGACCTGGTGTCTGAGTACCAGCAGTACCAGGACGCC ACTGCCGAGGAGGAGGGGGAGATGTATGAAGATGACGAC GAGGAGTCG	
GFAP		
Sense primer	5′-GACCTGCGACCTTGAGTCCT-3′	61.8
Antisense primer	5′-TCTCCTCCTTGAGGCTTTGG-3′	61.8
Probe	5′-FAM490-TCCTTGGAGAGGCAAATGCGC-BHQ-3′	68.9
Amplicon	GACCTGCGACCTTGAGTCCTTGCGCGGCACGAACGAGTCC TTGGAGAGGCAAATGCGCGAACAGGAGGAGCGCCACGCT CGGGAGTCGGCCAGTTACCAGGAGGCACTCGCTCGGCTG GAGGAGGAGGGCCAAAGCCTCAAGGAGGAGA	
MBP		
Sense primer	5′-ACTTGGCCACAGCAAGTACC-3′	61.8
Antisense primer	5′-GTGTGAGTCCTTGCCAGAGC-3′	62.0
Probe	5′-FAM490-TTGACTCCATCGGGCGCTTCT-BHQ-3′	68.3
Amplicon	ACTTGGCCACAGCAAGTACCATGGACCATGCCCGGCATGGC TTCCTCCCAAGGCACAGAGACACGGGCATCCTTGACTCC ATCGGGCGCTTCTTTAGCGGTGACAGGGGTGCGCCCAAG CGGGGCTCTGGCAAGGACTCACAC	
Nestin		
Sense primer	5′-GAGCTCTCTGGGCAAGTGGA-3′	63.0
Antisense primer	5′-CTCCCACCGCTGTTGATTTC-3′	62.9
Probe	5′-FAM490-AGGACAGTCAGCAGTGCCTGCA-BHQ-3′	67.8
Amplicon	GAGCTCTCTGGGCAAGTGGAACGTAGAGACCAGGGTAGA GGACAGTCAGCAGTGCCTGCAAGTAGAAGAGGGTCT GCAGGAGGAACAGCACCAAGAGTCTCTGAGAGAGGTG AAGCAGGAGCTGCCTAGCTCTGGAAATCAACAGCGGT GGGAG	
18S		
Sense primer	5′-GTAACCCGTTGAACCCCATTC-3′	62.6
Antisense primer	5′-CCATCCAATCGGTAGTAGCGA-3′	62.2

(continued)

Table 2 (continued)

Name	Sequence	T_M (°C)
Probe	5′-CAL610-AAGTGCGGGTCATAAGCTTGCG-BHQ-3′	67.6
Amplicon	GTAACCCGTTGAACCCCATTCGTGATGGGGATCGGGGATT GCAATTATTCCCCATGAACGAGGAATTCCCAGTAAGTGC GGGTCATAAGCTTGCGTTGATTAAGTCCCTGCCCTTTG TACACACCGCCCGTCGCTACTACCGATTGGATGG	

β-Tubulin III: neuronal marker; GFAP (glial fibrillary acidic protein): astrocytic marker; MBP (myelin basic protein): oligodendrocytic marker; Nestin: multipotent marker; 18S (18S ribosomal subunit): internal control; FAM490: FAM490 fluorophore; CAL610: CAL610 fluorophore; BHQ: Black Hole Quencher®; T_M: Melt temperature

25. PCR standard. Amplify the fragment of interest from recently prepared cDNA using the primers shown in Table 2. PCR clone this fragment into a plasmid. To do this, we recommend using either the TOPO-TA Cloning® Kit (Invitrogen) or the StrataClone™ PCR Cloning Kit (Stratagene, La Jolla, CA). Miniprep using a kit such as the Wizard Plus Minipreps DNA Purification System (Promega, Madison, WI), perform sequence analysis to confirm that the plasmid contains the correct amplicon (see Table 2), and measure A_{260} to quantify DNA concentration. Dilute to 1 ng/μl working stock and aliquot. Store at –20°C.

3. Methods

3.1. Culture, Cryo-Preservation, and Differentiation of Neural Progenitors

Adult hippocampal neural progenitors can be cultured *in vitro* for multiple passages and later cryopreserved for long-term storage. Morphologically, multipotent progenitors are very spherical, approximately 10–15 μm in diameter, phase-bright with a high nucleus to cytoplasmic volume ratio, and have only one or two short projections when viewed under a phase-contrast microscope at 40–100× (Fig. 1a). A culture is ready for passage when it is approximately 80% confluent (Fig. 1b), and cultures should be seeded at approximately 25,000 cell/cm^2. Therefore, the ideal passage protocol is a 1:4 subculture followed by 2 days of growth. Do not allow a culture to overgrow, as the cells will begin to differentiate and detach from the plate (unpublished observation).

Nearly any size culture dish is suitable for cell growth, provided that the surface is coated with poly-L-ornithine/laminin to allow cell attachment. We have successfully cultured cells in all formats ranging from 96-well plate to 10 cm culture dish,

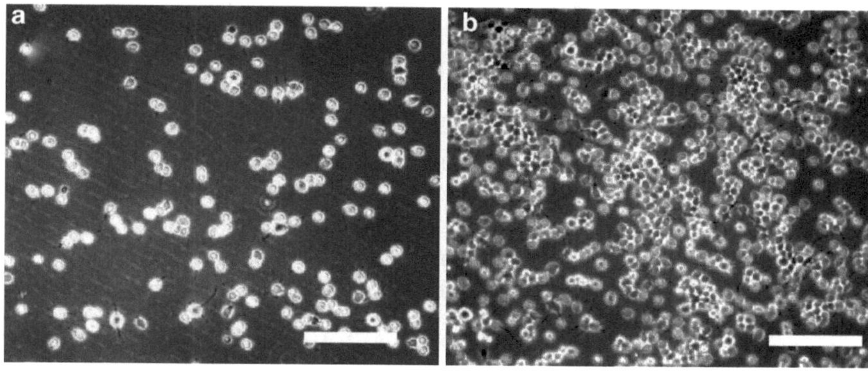

Fig. 1. Bright field images of adult neural hippocampal progenitors cultured in DMEM/F-12 + N-2 supplemented with 20 ng/ml FGF-2. (**a**) Approximately 1 h after plating on poly-L-ornithine/laminin coated dish. (**b**) Cells are ~80% confluent after 48 h in culture. Scale bars: 100 μm

including CultureSlides. However, we recommend using 3.5, 6, or 10 cm culture dishes for serial propagation of cells.

Once progenitors are isolated, avoid serum or antibiotics in culture to prevent differentiation and slow growth, respectively. Good aseptic technique is critical for the growth of AHNPCs in the absence of antibiotics. As with all cell culture, all open steps should be performed in a HEPA filtered biological safety cabinet by a person trained in good aseptic technique.

Progenitors are easily cryopreserved and generally recover quickly when thawed. The cells are best preserved in DMEM/F-12 + N-2 with 10% DMSO. As with most cell types, it is important to manipulate the cells gently when they are exposed to DMSO. Also, it is optimal to minimize the time that cells are exposed to DMSO in the liquid state, as DMSO will cause cells to differentiate (unpublished observation). Do not rush when freezing/thawing the cells, but do prepare all reagents and supplies (media warmed, vials labeled, etc.) before beginning those procedures.

3.1.1. Cell Culture

1. Examine the culture under a microscope to determine whether it is ready for passaging. The plate should be approximately 80% confluent (see Fig. 1b).
2. Warm DMEM/F-12 + N-2 to 37°C and thaw poly-L-ornithine/laminin coated plates. Plates can be thawed at either room temperature or 37°C, but avoid the formation of dry spots on the plate, as these will subsequently not effectively support cell attachment.
3. Aspirate all but ~0.5 ml of medium from the culture plate to prevent the plate from completely drying.
4. Add approximately 1 ml of cold Accutase per 75 cm^2 culture surface area. Never warm the Accutase stock and return it to 4°C immediately after use.

5. Incubate at 37°C for approximately 2 min to detach cells. Gently rock and shake the plate to fully dislodge progenitors. Do not bang the plate on a hard surface. Confirm under the microscope that all cells have detached (see Note 2).
6. Dilute the Accutase cell suspension in 5 ml of warm DMEM/F-12 + N-2. Gently wash the plate with this suspension 2–3 times to break up large cell clumps, then pipette into a 15 ml centrifuge tube.
7. Pellet cells at 200 g for 2 min (see Note 3).
8. Flame-polish a cotton-plugged Pasteur pipette (preferably inside a biological safety cabinet to maintain sterility of the full pipette). Gently rotate the tip of the pipette in the edge of a Bunsen burner flame until the flame turns yellow. Continue rotating the pipette and hold the tip in the flame for approximately 3 s. Rotating the pipette is crucial for achieving an even polish. After polishing, examine the tip. It should be smooth and evenly circular, and the diameter of the bore should be approximately half of its original size. Keep the polished pipette sterile by storing upright (in a tube rack, for example) in the biological safety cabinet.
9. Aspirate supernatant from the cell pellet, and add 2 ml DMEM/F12 + N-2. Other volumes may be used, but we recommend 2 ml for most applications.
10. Using the flame polished pipette, triturate the cell pellet by drawing nearly the whole 2 ml into the pipette and expelling it against the bottom of the centrifuge tube. Repeat 3–4 times. The pellet should be completely dissociated, and the solution should be turbid. Use care to prevent blowing bubbles when expelling the solution. Likewise, be careful to avoid drawing air into the pipette when aspirating (air–liquid interfaces promote cell lysis).
11. Optional: The number of cells in this suspension can now be quantified by hemacytometer counting. This is not necessary for routine cell maintenance; however, precise knowledge of cell concentration is required for experiments where inoculation density is critical.
12. Aspirate the PBS from the freshly thawed poly-L-ornithine/laminin coated dish, and immediately add warm DMEM/F-12 + N-2. Do not allow the plate to dry (see Note 4). Medium volume will depend on plate size and inoculum volume. Recommended culture volumes: 10 cm dish, 8 ml; 6 cm dish, 4 ml; 3.5 cm dish, 2 ml; 6-well plate, 2 ml; 12-well plate, 1 ml; 24-well plate, 0.5 ml; 96-well plate, 100 μl; 8-well CultureSlide, 500 μl.
13. Pipette cell suspension up/down several times to mix, because the cells settle very quickly. Then, add the appropriate volume

to the culture dish. Immediately after adding the inoculum, rock the plate back and forth several times to mix. To ensure a homogeneous inoculation, it is critical to thoroughly mix the plate immediately after adding the inoculum because the cells will settle and attach very quickly. Do not swirl the plate, as this will force the cells to edge of the plate.

14. Thaw an FGF-2 aliquot at room temperature and supplement the culture at 20 ng/ml (0.2 μl of FGF-2 stock per ml of culture). Gently rock the plate back and forth to mix.
15. Incubate at 37°C in 5% CO_2 atmosphere. Most cells should be attached within 30 min.

3.1.2. Cryopreservation

1. Follow Subheading 3.1.1 steps 1–10.
2. Determine the number of vials to be frozen. We recommend freezing approximately 2×10^6 cells per 1 ml vial. Therefore, each 80% confluent 10 cm dish will generate 4 vials. However, lower or higher cell concentrations can be used if desired.
3. Appropriately label all cryovials before continuing.
4. Prepare 20% DMSO freezing solution – 0.5 ml per vial plus 1 ml contingency.
5. Slowly and with constant and gentle mixing/swirling, add an equal volume of freezing solution to the cell suspension. Pipette up/down gently to mix.
6. Aliquot into cryovials at 1 ml per vial. If aliquotting a large number of vials, be sure to keep the cell suspension well-mixed.
7. Immediately transfer to –80°C. It is important to minimize the time that cells are exposed to liquid DMSO.
8. Store at –80°C at least overnight before transferring to liquid nitrogen storage. Cells can be stored at –80°C for up to 3 months without appreciable loss of viability, but longer-term storage requires liquid nitrogen.

3.1.3. Cell Thawing

1. Warm DMEM/F-12 + N-2 to 37°C and thaw a poly-L-ornithine/laminin coated plate. A standard vial frozen with 2×10^6 cells will require a 6 cm culture dish, but the dish size can be adjusted based on the application and number of cells. Plates can be thawed at either room temperature or 37°C.
2. Remove the vial from cryopreservation and immediately thaw in 37°C water bath with gentle swirling. Thawing will take 2–3 min (see Note 5).
3. Gently pipette the entire contents of the vial to a 15 ml centrifuge tube, taking care to avoid bubbles.
4. Slowly and with gentle mixing, add 5 ml warm DMEM/F-12 + N-2 to the cell suspension. Gently pipette up/down several times to mix.

5. Pellet cells at 200 g for 2 min.
6. Flame-polish a cotton plugged Pasteur pipette as in Subheading 3.1.1 step 8.
7. Aspirate supernatant and add 2 ml DMEM/F-12 + N-2 to the cell pellet.
8. Using the flame polished pipette, gently triturate the cell pellet as in Subheading 3.1.1, step 10.
9. Aspirate PBS from the freshly thawed poly-L-ornithine/laminin coated plate and immediately add warm DMEM/F-12 + N-2. Medium volume will depend on plate size (see Subheading 3.1.1 step 12).
10. Add the entire 2 ml of cell suspension to the plate. Immediately after adding the inoculum, gently rock the plate back and forth several times to mix.
11. Supplement the culture with 20 ng/ml FGF-2 (0.2 µl FGF-2 stock per ml of culture) and again rock the plate to mix.
12. Incubate overnight at 37°C in 5% CO_2 atmosphere.
13. To remove residual DMSO, aspirate and replace the medium with fresh, warm DMEM/F-12 + N-2 the day after inoculation. Resupplement with 20 ng/ml FGF-2.
14. Incubate at 37°C in 5% CO_2 atmosphere and monitor daily to determine when to pass the culture, which is generally the next day. Cells are ready for experimentation after one passage.

3.1.4. Differentiation

1. Follow the protocol in Subheading 3.1.1 to dissociate, count, and seed cells at 20,000 cells/cm^2. If, after differentiation, the culture will be analyzed by immunostaining (Subheading 3.2.2), we recommend using 8-well CultureSlides. If, after differentiation, the culture will be analyzed by quantitative RT-PCR (Subheading 3.2.3), we recommend using 3.5 cm dishes or 6-well culture plates (see Note 6).
2. Supplement the culture with 20 ng/ml FGF-2 and incubate overnight at 37°C in 5% CO_2 atmosphere.
3. Day 1: Remove FGF-2 medium and replace with differentiation medium of your choice (see Note 7):

Mixed population:	1 µM RA + 1% (v/v) FBS in DMEM/F-12 + N-2
Neuronal:	1 µM RA + 5 µM forskolin in DMEM/F-12 + N-2

Incubate at 37°C in 5% CO_2 atmosphere for 2 days.

4. Day 3: Replace media.

5. Day 5: Cultures are now ready for analysis. Proceed immediately to immunostaining protocol (Subheading 3.2.2) or quantitative RT-PCR protocol (Subheading 3.2.3).

3.2. Analyzing Culture State

3.2.1. Proliferation Assay

Quantifying the ability of neural progenitors to proliferate when exposed to variety of culture conditions and genetic manipulations is crucial for developing an understanding of the mechanisms governing the expansion of these cells. It is important to note, however, that this assay is not a measure of multipotency and simply measures cell number. Further experimentation is required to assess the lineage commitment capabilities of the cells (see Subheadings 3.2.2 and 3.2.3).

Proliferation can readily be measured with Cell Proliferation Reagent WST-1. This is a tetrazolium salt cleaved by mitochondrial dehydrogenase activity, producing formazan dye whose absorbance is quantified by spectrophotometry. After culturing cells in a 96-well plate format for 5 days, WST-1 is directly added to the culture, incubated for 2–3 h, and quantified on a microplate reader. Cell number is determined by comparison to a standard curve, where the linear range of absorbance vs. cell number is 1,000–50,000 cells per well.

Before beginning any experimentation, it is extremely important to plan the layout of the 96-well plate. We discourage using the edge wells for culturing the cells because evaporation over the 5-day culture period will affect the culture volume, which in turn will affect the spectrophotometric reading at the end of the experiment. The edge wells should be completely filled (300 μl) with sterile PBS until the final day of the experiment when they can be used for the blank and standard curve. We also recommend culturing each experimental condition in quintuplicate (5 wells) to ensure sound statistics. However, quadruplicate is acceptable. To help organize the 96-well plate, create a spreadsheet with a map of the plate, print it out, and refer to it often.

3.2.1.1. Day 0: Inoculation

1. Thaw 96-well poly-L-ornithine/laminin coated plate. This can be done at room temperature or 37°C.
2. Prepare all media solutions at 2× working concentration. For instance, prepare DMEM/F-12 + N-2 with 40 ng/ml FGF-2 if the experiment calls for 20 ng/ml FGF-2.
3. If not using the edge wells, which we do not recommend using, fill them completely (~300 μl) with sterile PBS.
4. Aspirate PBS from the other wells and immediately add 50 μl of 2× media. Do not allow the wells to dry out (see Note 4); therefore, this is best done by aspirating the PBS and adding media one row at a time. Place plate in 37°C incubator to warm the media.

5. Prepare cells. Follow Subheading 3.1.1 steps 1–11. Determine the cell concentration of the cell suspension by hemacytometer. This solution is generally quite concentrated, so a 1:10 dilution in DMEM/F-12 + N-2 prior to counting is usually necessary. This 1:10 dilution is generally sufficient for use in step 6. Keep cell suspensions well-mixed throughout the counting and diluting process.
6. Dilute the 1:10 solution used to quantify cell concentration to 20,000 cells/ml in DMEM/F-12 + N-2.
7. Using a multichannel pipette, add 50 µl of 20,000 cell/ml suspension to each well. To ensure that cells are evenly plated from one row to the next, the cell suspension in the pipette basin should be well mixed between each row. This can be achieved by pipetting up/down with the multichannel pipette multiple times.
8. Follow Subheading 3.1.1 steps 12–15 to continue culturing the leftover cells. They will be used on day 5 to generate a standard curve.
9. Incubate at 37°C in 5% CO_2 atmosphere. After 30 min, visually confirm under the microscope that cells are evenly plated across all wells.

3.2.1.2. Day 1–4: Media Changes

10. Perform daily 50% media changes on all wells. Using a multichannel pipette as appropriate, remove 50 µl of media and replace it with 50 µl fresh 1× media.
11. Continue passaging the leftover cells in parallel to give an ~80% confluent 10 cm dish on day 5.

3.2.1.3. Day 5: Plate Standard Curve and Quantify Cell Number by WST-1

12. Dissociate, dilute, and count the leftover cells maintained in parallel by following Subheading 3.1.1 steps 1–11.
13. Prepare standard curve and plate 100 µl per well of each cell concentration. Quadruplicate is best, but triplicate is sufficient. The cell concentrations can be varied based on the application, but we recommend 40,000, 20,000, 10,000, 5,000, 2,500, 1,250, and 0 (DMEM/F-12 + N-2 only) cells per well. There is no need to wait for cells to attach.
14. Thaw WST-1 at 37°C for at least 5 min. Briefly vortex to ensure all aggregates dissolve.
15. Using a multichannel pipette, add 10 µl WST-1 to each well. Gently rock plate to mix.
16. Incubate at 37°C for 2 h.
17. Rock on shaking platform for approximately 1 min.
18. Measure samples against blank (media only) on microplate reader. Measure absorbance at 440 and 700 nm. Subtract A_{700} from A_{440}. This is the experimental readout.

19. Plates can then be returned to the incubator and reanalyzed up to 4 h after WST-1 addition. Use the standard curve to determine the best read. This will depend on the cell concentrations of the standard curve, but we have generally found that 2–3 h post-WST-1 addition provides the best results.
20. Use the standard curve to interpolate cell numbers and perform statistical analyses as necessary.

3.2.2. Immunostaining

Immunostaining allows the researcher to directly visualize proteins expressed in a cell by probing with primary antibodies directed against the proteins with the appropriate secondary antibodies conjugated to fluorophores. This technique is typically used to determine a cell's lineage by staining for markers of neuronal (e.g., β-tubulin III or microtubule associated protein (MAP) 2a+2b), astrocytic (e.g., glial fibrillary acidic protein (GFAP)), oligodendrocytic (e.g., O4), and multipotent (nestin) cells. The resultant staining can be imaged and quantified to determine the percentage of cells positive for lineage markers of interest.

We recommend using a confocal microscope for generating high-quality images; however, a conventional fluorescent microscope is adequate to qualitatively assess the monolayer culture. Be sure to use the proper secondary antibodies and nuclear stain for your microscope. Because of the high associated costs, many confocal microscopes do not have a UV laser line, and therefore are unsuitable for imaging fluorophores that excite in the UV. This includes the common nuclear stain DAPI. Similarly, without the proper light-gathering hardware, it is impossible to visualize fluorophores that emit in the infrared with most fluorescent microscopes.

We recommend using the following fluorophores for secondary antibody staining: Alexa Fluor® 488 (green), Cy3 (red), and Cy5 (infrared). If nuclear staining is necessary, we recommend TO-PRO®3, which emits in the infrared and therefore should not be used simultaneously with Cy5. Before beginning staining, design the staining protocol and ensure that all of the proper primary and secondary antibodies are in stock. If costaining, do not use multiple primary antibodies raised in the same species and do not use fluorophores with similar emission spectra.

1. Culture the experiment on CultureSlides according to Subheading 3.1.1 and/or Subheading 3.1.4, depending on the experiment. The 8-well CultureSlides allow for multiple experimental conditions without sacrificing culture and staining convenience. However, other size CultureSlides may be used. Regardless of size, the CultureSlides must be coated with poly-L-ornithine/laminin (Subheading 2.1.1).
2. Fix cells with 0.5 ml of 4% paraformaldehyde per well for no longer than 10 min at room temperature (see Note 8).

This and all subsequent incubations should be done with gentle shaking, such as with an orbital shaker. If not immunostaining immediately, the cells can be stored in PBS at 4°C. If staining for O4, see Note 9 before proceeding.

3. Rinse three times with 0.3 ml PBS, 5 min per rinse at room temperature.
4. Block/permeablize for 30 min at room temperature with 0.3 ml PBS containing 5% donkey serum (or the appropriate normal serum) and 0.3% Triton X-100 (PBS-DT). (see Note 1).
5. Incubate with primary antibody diluted in PBS-DT (0.25 ml) for 24–48 h at 4°C. See Table 5.1 for recommended antibody dilutions.
6. Rinse twice with 0.5 ml PBS for 10 min at room temperature.
7. Rinse once with 0.5 ml PBS for 30 min at room temperature.
8. Incubate with 1:250 secondary antibody + 1:1000 TO-PRO®3 (1 μM) in PBS-DT (0.25 ml) for 2 h at room temperature. While incubating, cover with aluminum foil to protect from light. Care should be taken in all subsequent steps to minimize light exposure and avoid photobleaching.
9. Repeat steps 6–7.
10. Slides are now ready to be mounted. Remove gasket from the CultureSlide and allow sample to dry for 15 min.
11. Follow instructions in ProLong® Antifade Kit to prepare antifade/mounting medium. Pipette a line of ~200 μl antifade/mounting medium lengthwise along the edge of the slide, and slowly lay a coverslip over the slide, starting at the edge with the antifade/mounting medium. Proceed slowly and with care to avoid bubbles. Antifade/mounting medium should evenly coat the slide.
12. Dry the slide at room temperature in the dark for at least 2 h (overnight drying is acceptable). Seal the slide with clear nail varnish. Apply the varnish carefully to avoid covering the samples. Dry at room temperature in the dark for approximately 30 min.
13. Slides are now ready to be imaged and can be stored in the dark at 4°C. Alexa Fluor® 488 excites at 495 nm and emits at 519 nm (green). Cy3 excites at 550 nm and emits at 570 nm (red). Cy5 excites at 650 nm and emits at 670 nm (infrared). TO-PRO®-3 nuclear stain excites at 642 nm and emits at 661 nm (infrared). See Fig. 2 for typical β-tubulin III, MAP 2a + 2b, GFAP, O4, and nestin staining images.

3.2.3. Quantitative RT-PCR

Quantitative reverse transcription (RT)-PCR is a useful tool for determining the state of a cell population by measuring the concentration of a specific mRNA transcript. It is important to

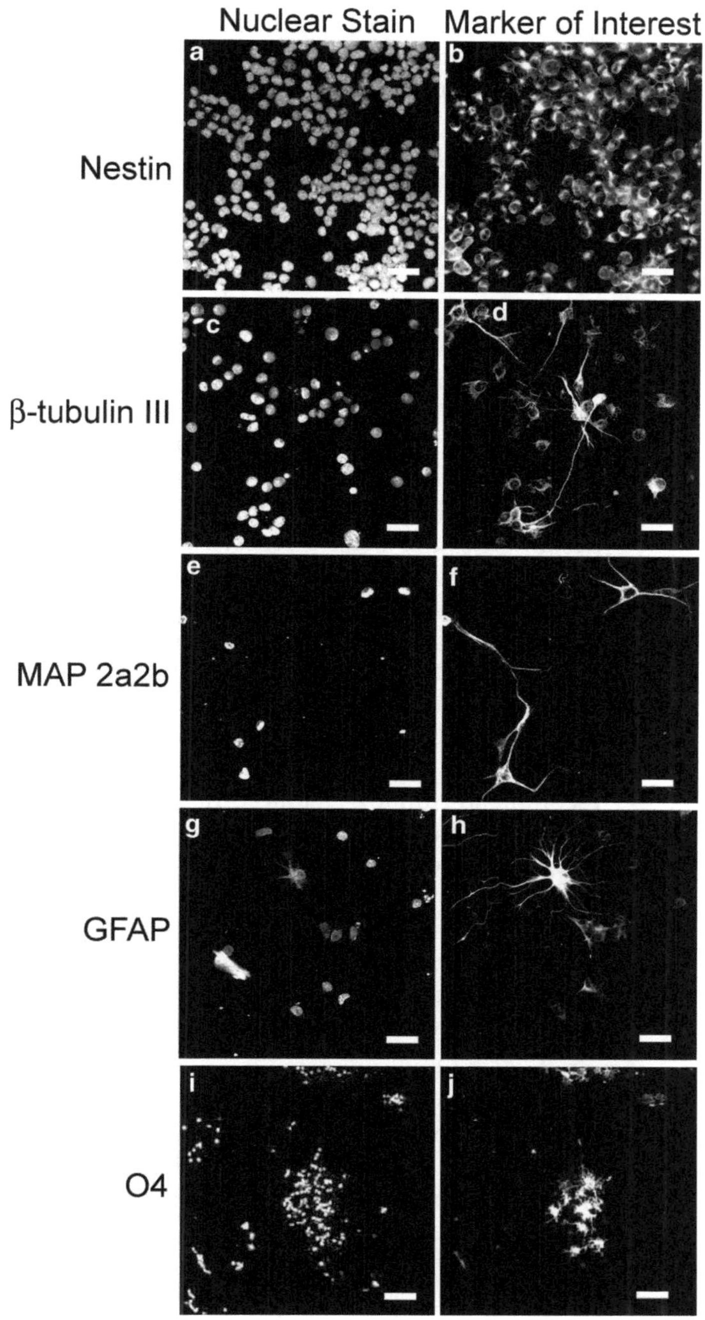

Fig. 2. Representative immunostaining of (**b**) the multipotent marker nestin, (**d**) neuronal marker β-tubulin III, (**f**) neuronal marker MAP 2a + 2b, (**h**) astrocytic marker GFAP, and (**j**) oligodendrocytic marker O4. (**a, b**) Cultures were grown in DMEM/F-12 + N-2 supplemented with 20 ng/ml FGF-2 for 2 days and stained for nestin. (**c, d**) Cultures were grown in neuronal differentiation medium (DMEM/F-12 + N-2 supplemented with 1 μM RA + 5 μM forskolin) for 5 days and stained for β-tubulin III. (**e–j**) Cultures were grown in mixed population differentiation medium (DMEM/F-12 + N-2 supplemented with 1 μM RA + 1% FBS) for 4 days and stained for MAP 2a + 2b, GFAP, or O4, respectively. (**a**, **c**, **e**, **g**, **i**) TO-PRO®-3 nuclear stain. (**b**, **d**, **f**, **h**, **j**) marker of interest. Scale bars: (**a–h**) 16 μm, (**i**, **j**) 50 μm

note that quantitative RT-PCR measures the population-averaged concentration of mRNA for an entire culture. If cell-to-cell variability within a culture is important, then we recommend analyzing the culture via immunostaining to gain single cell data (Subheading 3.2.2). We have demonstrated the utility of quantitative RT-PCR to quantify the cell differentiation state through the measurement of neuronal (β-tubulin III), astrocytic (GFAP), oligodendrocytic (myelin basic protein (MBP)), and multipotent (nestin) markers (5). Here, we present a method for quantifying expression of these marker genes in neural progenitors; however, we do not cover the design and testing of new primer and probe sets. We refer the reader to standard references for a detailed explanation of this procedure (6).

Quantitative RT-PCR is accomplished in four steps: cell culture, RNA isolation, reverse transcription, and quantitative PCR. The cell culture conditions will depend on the experimental design and have been discussed previously (Subheading 3.1). To ensure sound statistics, we recommend culturing each experimental condition in triplicate. RNA isolation is accomplished with TRIzol® reagent, and this is followed immediately by reverse transcription. Both of these methods draw heavily from the manufacturer's (Invitrogen) instructions. The resulting cDNAs are quantified by Taqman probe QPCR using the 18S ribosomal subunit as an internal control. Again for statistical purposes, we recommend running technical triplicates for each cDNA sample. Combined with the biological triplicate, this results in nine QPCR reactions per condition.

3.2.3.1. RNA Isolation

1. The RNases that exist on most surfaces easily degrade RNA. Therefore, it is extremely important to use only RNase-free reagents and supplies and to wash all equipment, surfaces, and gloved hands with RNase Away before beginning work.
2. Aspirate media from cells, wash once with PBS, and add TRIzol®. The ideal TRIzol® volume for sample processing in micro-centrifuge tubes is 1 ml. This volume is adequate to cover 3.5 cm, 6 cm, and 6-well plates; however, this and all subsequent steps must be scaled appropriately for larger plates.
3. Scrape the surface of the plate with a cell lifter to detach all cells. Then, triturate with 1 ml Pipetman to dissociate clumps.
4. Pipette TRIzol® solution into micro-centrifuge tubes (see Note 10) and incubate at room temperature for 5 min. Samples can now be stored long-term at −80°C and slowly thawed on ice for further processing. Beyond this point, samples should be kept on ice at all times unless otherwise noted.
5. Add 200 μl chloroform and vortex vigorously for 15 s.
6. Let samples phase separate on ice for 2–3 min. Spin at 12,000 g for 15 min at 4°C.

7. Pipette the upper, aqueous phase to a fresh, RNAse-free, micro-centrifuge tube. The interface between the aqueous and organic phases contains cellular proteins. Do not disturb this interface when pipetting, or protein contamination may result.
8. Add 1 μl DNase and incubate at room temperature for 15 min.
9. Repeat steps 5–7 to remove DNase and residual TRIzol®.
10. Add 5 μl of glycogen. This will improve visibility of the RNA pellet and increase yield.
11. Add 500 μl of isopropanol and incubate at room temperature for 10 min (see Note 11).
12. Centrifuge at 12,000*g* for 10 min at 4°C.
13. Pour off the supernatant and wash the pellet with 1 ml 75% ethanol. Spin at 7,500 g for 5 min at 4°C.
14. Pipette off the supernatant (this pellet is looser) and briefly air-dry the pellet for 5–10 min at room temperature. Do not overdry.
15. Resuspend the pellet in 20 μl RNase-free water.
16. Proceed immediately to reverse transcription. Alternatively, RNA can be stored at –80°C and thawed on ice for later processing.

3.2.3.2. Reverse Transcription

17. Continue to use care to prevent RNase contamination (see step 1) until after the reverse transcription reaction is complete.
18. Quantify RNA concentration by measuring A_{260} on a spectrophotometer. Predilute samples in RNase-free water if necessary, depending on the spectrophotometer. A 3.5 cm plate of cells should yield roughly 20–40 μg of RNA, depending on experimental conditions (see Note 12).
19. For each sample, mix the following in a PCR (0.2 ml) tube:

Random hexamer primer:	1 μL
10 mM dNTP mix:	2 μL
RNA (10 pg – 5 μg):	*x* μL
DEPC – water:	to 12 μL

The volume of RNA will depend on the concentration of the sample. We recommend normalizing all samples within an experiment such that each reverse transcription reaction has the same amount of RNA. Each reaction should have 10 pg to 5 μg RNA; however, it is best to normalize all reactions to the highest concentration possible.

20. Denature this mixture by incubating in a thermal cycler at 65°C for 5 min. Immediately place on ice (see Note 13).
21. Prepare the following master reaction mix on ice. Volumes shown below are per reaction. Make approximately 10% more master reaction mix than needed as a contingency. Vortex to mix.

5× cDNA synthesis buffer:	4 µl (vortex before use)
0.1 M DTT:	1 µl
RNase OUT™:	1 µl
DEPC water:	1 µl
ThermoScript™ RT:	1 µl (keep on ice at all times)

22. Add 8 µl master reaction mix to each PCR tube. Keep on ice.
23. Perform the reverse transcription reaction by incubating in thermal cycler.

    25°C 10 min

    55°C 50 min

    85°C 5 min
24. To digest the remaining RNA templates, add 1 µl RNase H to each reaction and incubate at 37°C for 20 min (see Note 14). DO NOT use RNase-free supplies for this step, as they will become contaminated with RNase.
25. The resulting cDNAs can be used immediately for quantitative PCR or stored at −20°C for up to 2 weeks. Quantitative data are unreliable from cDNAs older than 2 weeks; however, these samples are adequate for semi-quantitative experiments or cDNA amplification for cloning.

3.2.3.3. QPCR

26. Prepare standard curve in micro-centrifuge tubes. Mix 10 µl of 1 ng/µl gene-of-interest standard with 10 µl of 1 ng/µl 18S ribosomal subunit standard. Using distilled/de-ionized water, serially dilute in tenfold increments to generate a standard curve spanning 7 orders of magnitude.
27. Add 6 µl of each cDNA sample and each standard curve sample to individual micro-centrifuge tubes.
28. Prepare PCR reaction mix. See Table 2 for primer and probe sequences. Mix the following components in the order shown. Volumes shown are per 20 µl PCR reaction. As a contingency, make approximately 10% more reaction mix than needed. Vortex to mix.

2× Supermix	10 µl
Distilled/de-ionized water	7 µl
Gene-of-interest sense primer(10 µM stock)	0.5 µl
Gene-of-interest antisense primer (10 µM stock)	0.5 µl
18S sense primer (10 µM stock)	0.5 µl
18S antisense primer (10 µM stock)	0.5 µl
Gene-of-interest probe (10 µM stock)	0.4 µl
18S probe (10 µM stock)	0.4 µl
Antibody/Taq polymerase	0.25 µl

Table 3
Quantitative PCR protocol

Cycle	Repeats	Step	Time (mm:ss)	Setpoint (°C)
1	1	1	05:00	95
2	50	1	00:30	95
		2[a]	00:30	58
		3	00:15	72

[a]Indicates the real-time detection step

29. Add 60 µl PCR reaction mix to each sample. Pipette up/down to mix.
30. Transfer 20 µl to each well on a 96-well PCR Plate. Plate each sample and standard in triplicate. The resulting standard curve now has concentrations from 1 ng/well to 10^{-7} ng/well of both the 18S and gene-of-interest standards.
31. Seal the plate with the Microseal B Adhesive Seal. Run a hard, clean, straight edge across the top of the plate to create a strong seal around each well.
32. Transfer the plate to real-time PCR detection system, and run the protocol shown in Table 3. Be sure to select the correct fluorophores (FAM-490 for gene-of-interest and CAL-610 for 18S ribosomal subunit – the Texas Red setting may be substituted for CAL-610).
33. PCR efficiencies should be near 100% (±10%). Construct standard curves for the gene-of-interest and 18S ribosomal subunit and interpolate unknown sample values. Normalize gene-of-interest values to 18S ribosomal subunit and perform appropriate statistical analyses (see Note 15).

4. Notes

1. Normal serum is serum derived from the species in which the secondary antibody was raised. Use of nonmatched secondary antibodies and blocking serum can lead to nonspecific, background staining.
2. Differentiated cells are much more adherent than multipotent progenitors. If, after several minutes of Accutase incubation and gentle rocking, there are large areas of attached cells, do not attempt to forcibly remove them. Most likely, they are lineage committed and not desirable for further passaging.
3. The medium does not contain serum; therefore, it will not inactivate the Accutase. The Accutase must be removed by centrifugation for cell pelleting followed by aspiration of the Accutase-containing medium.
4. Always add media to the plate before adding cells to prevent drying, as the dry regions will not support cell attachment.
5. If the vial was stored in liquid nitrogen, proceed immediately to thawing. Do not transfer to −80°C or dry ice for any longer than a few minutes. This will result in extremely poor viability after thaw.
6. If a quantitative RT-PCR experiment involves a time-course study, then do not use 6-well plates. RNA extraction requires lysing cells directly from the plate using the highly toxic TRIzol® reagent. Residual TRIzol® in a well, which contains volatile components, is sufficient to kill cells in neighboring wells.
7. If a long-term (10–14 days) differentiation experiment is required, use DMEM/F-12 + N-2 supplemented with 1 ng/ml FGF-2 + 10 ng/ml neurotrophic factor 3 (NT-3, Peprotech) + 10 ng/ml brain derived neurotrophic factor (BDNF, Peprotech). Change media every other day.
8. All volumes in the immunostaining protocol are for 8-well CultureSlides. Other sizes may be used, but volumes must be adjusted accordingly.
9. If immunostaining for O4, do not permeablize with Triton X-100. O4 is a surface marker which can be disrupted by Triton X-100. For *in vitro* staining, follow Subheading 3.2.2 steps 1–9 without Triton X-100, postfix with 4% paraformaldehyde, and then continue with permeablization and costaining at step 3.
10. TRIzol® already contains RNase inhibitors, so there is no need to use RNase-free tubes for this step.
11. As a convenient stopping point, samples may be incubated overnight on ice at 4°C.

12. To verify the integrity of the isolated RNA, the ratio A_{260}/A_{280} should be 1.9–2.1. Additionally, a small aliquot of RNA can be analyzed on a 1% agarose gel and visualized with ethidium bromide. The predominant bands are 18S and 28S rRNA at approximately 1 and 2 kilobases, respectively, when compared to a DNA ladder.
13. Set the thermal cycler to automatically drop to 4°C after denaturation. This avoids accidentally overheating the samples at the end of the run.
14. RNase H cleaves RNA phosphodiester bonds in DNA:RNA duplexes. Also note that RNase H is not inhibited by RNase OUT™.
15. The purpose of the standard curve is only to provide a reference on which to base experimental values. The experimental "concentrations" obtained from interpolating the standard curve do not correspond to the absolute quantities of mRNA in the culture. These concentrations provide relative values for comparing data from multiple experiments.
16. The mouse anti-O4 antibody is an IgM. Confirm that the intended secondary antibody will recognize IgM before use.

Acknowledgments

This work was funded by an NSF Graduate Research Fellowship (to JP), a California Institute for Regenerative Medicine Training Grant (T1-00007), NIH T32 GM007352 (to MJR), and NSF BES-0629202.

References

1. Ray J, Peterson DA, Schinstine M, Gage FH (1993) Proliferation, differentiation, and long-term culture of primary hippocampal neurons. Proc Natl Acad Sci USA 90:3602–3606
2. Palmer TD, Takahashi J, Gage FH (1997) The adult rat hippocampus contains primordial neural stem cells. Mol Cell Neurosci 8:389–404
3. Lai K, Kaspar BK, Gage FH, Schaffer DV (2003) Sonic hedgehog regulates adult neural progenitor proliferation in vitro and in vivo. Nat Neurosci 6:21–27
4. Peltier J, O'Neill A, Schaffer DV (2007) PI3K/Akt and CREB regulate adult neural hippocampal progenitor proliferation and differentiation. Dev Neurobiol 67:1348–1361
5. Abranches E, O'Neill A, Robertson MJ, Schaffer DV, Cabral JM (2006) Development of quantitative PCR methods to analyse neural progenitor cell culture state. Biotechnol Appl Biochem 44:1–8
6. Bookout AL, Cummins CL, Kramer MF, Pesola JM, Mangelsdorf DJ (2006) High-throughput real-time quantitative reverse transcription PCR. In: Ausubel FM, Brent R, Kingston RE, Moore DD, Seidman JG, Smith JA, Struhl K (eds) Current protocols in molecular biology, Wiley, New York, Unit 15.8

Chapter 6

In Vivo Analysis of Engrafted Adult Hippocampal Neural Progenitors

Matthew J. Robertson, Joseph Peltier, and David V. Schaffer

Abstract

A neural degenerative disease is characterized by the deterioration of neural tissue and subsequent loss of function. The *in vivo* engraftment of neural stem cells is a promising approach to the functional replacement of neural tissue with the ultimate goal of regaining lost function. In addition, by studying the behavior of engrafted neural stem cells in healthy and diseased tissue, insight can be gained into the extracellular and intracellular mechanisms which regulate stem cell behavior *in vivo*. Adult hippocampal neural progenitor cells (AHNPCs) are one potential source of cells that can be used to this goal. In this chapter, we describe some of the *in vivo* techniques necessary to study hippocampal progenitors in the adult rat, including engraftment and analysis by immunofluorescent staining. These techniques are important for studying AHNPCs within the physiologically relevant environment of the adult brain rather than in a culture dish.

Key words: Neural, Hippocampus, Progenitors, Rat, *In vivo*, Adult, Engraftment, Immunostaining, Pulse-chase/BrdU, Neurons, Astrocytes, Differentiation

1. Introduction

Hippocampal progenitors have been successfully engrafted into numerous regions of the nervous system, including the two constitutively active neurogenic regions of the adult brain: the hippocampus and the subventricular zone to olfactory bulb pathway. After engraftment into both regions, cells have differentiated and expressed markers consistent with neurons native to the engraftment site (1, 2). Grafting hippocampal progenitors provides a means to determine their plasticity and identify new regulatory signals when the mechanisms that control their behavior are incompletely understood, in either healthy or damaged/diseased tissue. Furthermore, their ability to

Irina M. Conboy et al. (eds.), *Protocols for Adult Stem Cells*, Methods in Molecular Biology, vol. 621,
DOI 10.1007/978-1-60761-063-2_6, © Springer Science+Business Media, LLC 2010

differentiate and integrate into the nervous system provides hope for potential cell-based therapies for the treatment of neurodegenerative diseases (3).

Grafting studies involve first marking the cells for tracking postengraftment. For simplicity we will discuss using BrdU to mark cells, but other methods are available (see Note 1). After the cells have been marked, they can either be cultured under conditions that induce differentiation prior to engraftment (see Chapter 5, Subheading 3.1.4 for common conditions), or grown in proliferative culture conditions (20-ng/mL FGF-2). Before proceeding, we assume the reader is familiar with the growth and differentiation techniques outlined in Chapter 5. After engraftment, the cells are allowed to acclimate and integrate into the host, from several weeks to several months (1, 2). Longer time points allow the cells to fully differentiate and to functionally integrate. Lastly, the brains are harvested and examined for colocalization of BrdU and neural lineage markers to determine the progenitor cell fate.

2. Materials

2.1. Preparation of Samples for Grafting

1. Dulbecco's Modified Eagle Medium/Nutrient Mix F-12 (DMEM/F-12) with HEPES and L-Glutamine (Invitrogen, Bethesda, MD) supplemented with 1% (v/v) N-2 Supplement (Invitrogen). Store supplemented medium at 4°C for up to 1 month.
2. Accutase (Innovative Cell Technologies, San Diego, CA).
3. Basic fibroblast growth factor (FGF-2, Peprotech, Rocky Hill, NJ). FGF-2 stock solution: dissolve in DMEM/F-12 + N-2 at 0.1 mg/mL, aliquot, and store at –20°C. Once thawed, aliquot can be stored at 4°C for up to 1 week.
4. Sterile cotton-plugged glass Pasteur pipettes.
5. 5-Bromo-2′-deoxyuridine (BrdU, Sigma, St. Louis, MO). Prepare 5 mM stock solution in distilled/deionized water and 0.22 μm syringe filter.
6. Phosphate buffered saline (PBS), pH 7.4 (Invitrogen).
7. Poly-L-ornithine/laminin-coated culture plates (see Subheading 2.1.1).

2.1.1. Laminin-Coated Culture Plates

Reagents

1. Sterile tissue culture water.
2. Sterile PBS, pH 7.4 (Invitrogen, Bethesda, MD).

3. Poly-L-ornithine hydrobromide (Sigma) dissolved in tissue culture water at 10 mg/mL. Sterilize by 0.22 μm syringe filtration, aliquot, and store at −20°C. Avoid repeated freeze/thaw.
4. Natural mouse laminin (Invitrogen). Store at −80°C. Thaw slowly at 4°C before use. Avoid repeated freeze/thaw and vortexing.
5. Plastic wrap.

Preparation of Plates

6. Dilute poly-L-ornithine in tissue culture water to a final concentration of 10 μg/mL and add to tissue culture plates such that plates are well covered (~7 mL for 10 cm plate). Incubate overnight at room temperature. To maintain sterility, leave the plates in a biological safety cabinet with blower off and sash closed. Leave UV light OFF.
7. Rinse plates twice with approximately the same amount of sterile water as used for poly-L-ornithine incubation. Do not allow plates to dry out.
8. Dilute laminin in sterile PBS to a final concentration of 5 μg/mL and add to plates at the same volume as was used for poly-L-ornithine.
9. Incubate at 37°C overnight.
10. Wrap in plastic wrap and store at −20°C for up to 6 months. Freeze plates on a level surface to avoid pooling and dry spots.

2.2. Surgical Grafting and Tissue Sectioning

1. Puralube Vet Ointment (PharmaDerm, Duluth, GA).
2. Sterotaxic instrument (e.g., Benchmark™ Stereotaxic Instrument, myNeuroLab.com, St. Louis, MO).
3. Betadine Vet Solution (Fisher, Pittsburgh, PA).
4. Scalpel, No. 10 blade.
5. 5 μL Hamilton syringe.
6. 26-gauge beveled needle.
7. Dental drill for surgery (Dremel, Racine, WI).
8. Surgical staples.
9. 70% Ethanol.
10. PBS, pH 7.5 (Invitrogen).
11. 0.9% (w/v) NaCl, approximately 150 mL per animal (0.22 μm sterile filtered).
12. Institutionally approved analgesia (e.g., 0.05 mg/kg buprenorphine dissolved in sterile, warm 0.9% NaCl).

13. Peristaltic pump for perfusion (e.g., Carter Manostat, Fisher).
14. Tubing for perfusion, 2.79 mm inner diameter (Fisher).
15. Blunted 17- or 18-gauge needle (or needle hand-pump attachment).
16. Surgical scissors.
17. Large straight forceps.
18. Two small curved forceps.
19. Rongeurs forceps.
20. Spatula.
21. Ice cold 4% (w/v) paraformaldehyde (PFA). While stirring, carefully heat 400 mL water on a hot plate in a fume hood to 60°C. Do not heat above 60°C, because aldehyde will form and cause autofluorescence. While maintaining temperature, add 20 g PFA and stir. As necessary, add NaOH pellets one at a time until paraformaldehyde dissolves. Generally, 1–2 NaOH pellets over the course of 10–15 min will fully dissolve the PFA. There will still be small particulates in suspension. Cool to room temperature. Add 50 mL 10× PBS and pH to 7.4 with HCl. Bring total volume to 500 mL with water and sterile filter to remove particulates. Prepare approximately 300 mL per animal. During perfusions keep the PFA on ice. 4% PFA must be disposed as chemical waste.
22. 20 mL scintillation vials (Wheaton, Millville, NJ).
23. 30% sucrose. Dissolve 30 g of sucrose in sufficient water for 100 mL of final solution. Sterilize the solution with a 0.22 μm filter and store at 4°C.
24. Tissue cryoprotectant: 25% ethylene glycol, 25% glycerol, 50% 0.1 M Na_2HPO_4, pH 7.4. Store at 4°C.
25. 96-well plate for storing sections.
26. Parafilm.
27. Sliding microtome (e.g., Vibratome, St. Louis, MO).
28. Trypan blue stain (Invitrogen).
29. Hemacytometer.

2.2.1. Anesthesia

1. Ketamine Hydrochloride, 100 mg/mL (Phoenix Scientific, St. Joseph, MO).
2. Xylazine Hydrochloride, 20 mg/mL (Ben Venue Laboratories, Bedford, OH).
3. 9% (w/v) NaCl in distilled/de-ionized water. Sterilize by 0.22 μm filtration.
4. Anesthetic cocktail: For each 150 g rat, sterilely mix 135 μl ketamine, 75 μl xylazine, 50 μl 9% NaCl, and 240 μl distilled/de-ionized water.

Table 1
Recommended primary antibodies, dilutions, and storage conditions for immunostaining

Name	Species	Supplier	Dilution	Storage
β-Tubulin III	Mouse	Sigma	1:250	Aliquot –20°C
MAP2a+2b	Mouse	Sigma	1:250	Aliquot –20°C
GFAP	Rabbit	Abcam (Cambridge, MA)	1:500	4°C
O4 (see Note 9)	Mouse	Millipore (Billerica, MA)	1:250	4°C
Nestin	Mouse	BD Pharmingen (San Jose, CA)	1:1,000	4°C
BrdU	Mouse	Roche Applied Science (Indianapolis, IN)	1:40	Aliquot –20°C=

2.3. Tissue Section Immunostaining

1. 12- and 24-well plates.
2. Cell culture inserts (net carriers) for 12-well plates (BD Falcon, Bedford, MA).
3. Paint brush, fine point (for transferring and manipulating sections).
4. Tris buffered saline (TBS): Add 6.05 g Tris base and 8.76 g NaCl to 900 mL water. pH to 7.5 with HCl and bring up to 1 L with water. Store at room temperature.
5. Triton X-100 (EMD Biosciences, San Diego, CA).
6. Normal serum (see Note 2).
7. Primary antibodies (see Table 1).
8. Cy3 and Cy5 conjugated donkey anti-rabbit and anti-mouse antibodies (Jackson ImmunoResearch, West Grove, PA). Upon receipt, reconstitute in 1:1 glycerol:water according to package insert (volume varies by lot), aliquot, and store at –20°C protected from light. Avoid freeze/thaw.
9. TO-PRO®-3 nuclear stain (Invitrogen). Upon receipt, aliquot and store at –20°C. Protect from light and avoid freeze/thaw.
10. Alexa Fluor® 488 conjugated donkey anti-mouse and anti-rabbit antibodies (Invitrogen). Upon receipt, aliquot and store at –20°C protected from light. Avoid freeze/thaw.
11. 4% (w/v) PFA (see Subheading 2.2, item 21 for preparation instructions).
12. 10× SSC: 3 M sodium chloride and 0.3 M sodium citrate in water. pH to 7.0. Make 50 mL.

13. 50% Formamide solution: Mix 50 mL of formamide in 10 mL of 20× SSC solution and 40 mL of deionized water. Pre-made 50% formamide solution can be stored at −20°C.
14. 2-N hydrochloric acid. Make 50 mL.
15. 0.1 M borate buffer: Add 3.2 g boric acid to 400 mL water. pH to 8.5 with NaOH and bring to 500 mL with water. Store at room temperature.
16. Gelatin-coated slides, 3″ × 1″ (LabScientific, Livingston, NJ).
17. Microscope cover glasses (24 × 60 mm, Fisher).
18. ProLong® Antifade Kit (Invitrogen).

3. Methods

3.1. Preparation of Samples for Engraftment

1. One day prior to engraftment, plate cells as described in Chapter 5, Subheading 3.1.1 and supplement the cell culture media with BrdU to a final concentration of 5 μM (1 μL of a 5 mM BrdU stock per mL of culture). Plate enough cells to ensure approximately 10^5 cells per graft (for reference, an 80% confluent 10 cm plate will have approximately 8×10^6 cells).
2. Incubate the cells in DMEM/F-12 + N-2 + 20 ng/mL FGF-2 + 5 μM BrdU at 37°C in 5% CO_2 atmosphere for 12 h.
3. Aspirate the medium, leaving behind enough so the culture does not dry out. Detach the cells and suspend as described in Chapter 5, Subheading 3.1.1, steps 1–10.
4. Pellet the cell suspension at 200 g for 2 min, and prepare another flame polished pipette (see Chapter 5, Subheading 3.1.1, step 8).
5. Aspirate the supernatant from the cell pellet and add 5 mL of PBS (see Note 3).
6. Suspend the cell pellet as described in Chapter 5, Subheading 3.1.1, step 10.
7. Repeat steps 4–6, but suspend the cells in only 1 mL of PBS.
8. Transfer the cell suspension to a 1.5 mL centrifuge tube.
9. Using a hemacytometer or other preferred means of cell quantification determine the number of cells in the suspension.
10. Pellet the cell suspension at 200 g for 2 min.
11. Carefully aspirate the supernatant from the cell pellet and add sufficient PBS to yield a final concentration of 75,000–100,000 cells per μL. Use a Pipetman for aspiration to avoid disturbing the pellet. The actual cell concentration

will be lower than the calculated due to residual liquid on the pellet. Ideally, by targeting a higher concentration the final concentration will be approximately 50,000 cells per μL.

12. Take 3 μL of the final cell suspension and dilute 1:10 in PBS. Using the 1:10 dilution, determine the cell concentration in the final suspension. Store the remaining 20 μL on ice for determining cell viability (see Subheading 3.2, steps 17–22).
13. Using a Pipetman, carefully measure the final volume of the cell suspension. Before starting the animal procedure, ensure there is sufficient cell suspension for all the planned engraftments. It is best to have sufficient cell suspension for two to five extra engraftments as a contingency.
14. Close the centrifuge tube and place the cell suspension on ice to maintain cell viability.

3.2. Grafting

This protocol is a general description of how to perform a stereotaxic graft of hippocampal progenitors into an adult rat's brain. To assess cell engraftment, the rat is perfused transcardially so that the cells can be analyzed histologically at time-points after the graft that would permit assessment of proliferation (short survival) and differentiation (longer survival – typically 4 weeks or more).

Similar to engraftment, endogenous hippocampal progenitors can be virally transduced by stereotaxic injection of a retrovirus produced as described in Chapter 7. This virus may contain a gene that will simply mark the progenitors (e.g., GFP) or it may contain some other gene of interest. Rather than injecting cultured progenitors as described in this section, approximately 2–3 μL of purified virus is injected.

Anesthesia

1. Anesthetize the rats using ketamine/xylazine anesthetic cocktail (500 μL of anesthesia per 150 g of body weight) via intraperitoneal injection.
2. Wait several minutes to ensure that the rat is anesthetized. Check reflexes by pulling the hind limb laterally and pinching the hind paw while the animal is on its stomach. If the rat is not fully under it will retract its hind paw.
3. When there is no withdrawal reflex, apply a small drop of Puralube Vet Ointment to each eye to prevent dryness, as the animals are unable to blink while under anesthesia. Proceed to the sterotaxic injection.

Stereotaxic Surgery

4. Assemble the sterotaxic apparatus with the 5 μL Hamilton syringe attached to a 26-gauge beveled needle.

5. Shave the hair on the top of the animal's head.
6. Mount the anesthetized rat in the sterotaxic apparatus, using ear bars and nose bar, as per the manufacturer's directions to ensure the skull is level (bregma and lambda are aligned horizontally). Sterilize the shaved skin with an approved antiseptic such as Betadine.
7. Using a scalpel, make a medial incision stretching from bregma to lambda. Retract the skin and underlying membranes to expose the skull. Align the needle over bregma, and note the coordinates on the stereotaxic instrument. Using a rat atlas such as Paxinos and Watson (4) to determine coordinates for the desired injection site, move the needle to the desired location over the skull, and use the drill to slowly drill through the skull being careful not to disturb the underlying dura or brain tissue (see Note 4).
8. Raise the Hamilton syringe to its highest possible position in the stereotaxic instrument, place the tube containing cells underneath it, and use the Hamilton syringe to gently mix the prepared cells from Subheading 3.1, step 14.
9. Using the syringe, measure sufficient cell suspension for 75,000 cells based on the concentration calculated in Subheading 3.1, step 12 (e.g., for 50,000 cell/μL measure 1.5 μL).
10. Slowly lower the needle to the correct dorsal/ventral sterotaxic coordinate.
11. As smoothly as possible, inject liquid at the rate of 1 μL/min until complete. An infusion pump can be used to ensure a slow, even injection.
12. Raise the syringe 1 mm, wait 1 min, and then raise the syringe slowly until it exits the brain.
13. Remove the rat from the sterotaxic apparatus and close the incision using surgical staples.
14. Subcutaneously inject 1 mL of 0.9% warm NaCl to hydrate the animal and administer your institutionally accepted analgesia (e.g., 0.05 mg/kg buprenorphine dissolved in sterile, warm 0.9% NaCl).
15. Between engraftments rinse the syringe and needle first with PBS, then with 70% ethanol, and finally rinse three times with PBS. Return to step 1 for the next animal.
16. Wait for the cells to exhibit the desired behavior (proliferation or differentiation), often over a period of weeks, then proceed to perfusion.

Cell Viability

17. To ensure any differences observed after engraftment are not due to cell preparation and handling, it is best to perform a

cellular viability assay (e.g., trypan blue exclusion) before and after the animal surgeries using any extra cells. For this purpose, it is sufficient to use the 1:10 diluted cells from Subheading 3.1, step 12 (see Note 5).

18. Using the 1:10 dilution of cells from Subheading 3.1, step 12 or other appropriate source of cells, transfer 10 μL of suspension to a fresh 1.5 mL microcentrifuge tube.
19. Add 10 μL of trypan blue to the fresh aliquot and mix gently 2–3 times (one part trypan blue solution per one part cell suspension).
20. Incubate at room temperature for 3 min (see Note 6).
21. Using a hemacytometer count the unstained (viable) and stained (nonviable) cells.
22. The fraction of viable cells is calculated as the number of viable cells (unstained cells) divided by the total number of cells (unstained cells + stained cells). Ideally this fraction will be >90% and similar for each sample used for engraftment before and after the animal surgery.

Perfusion

23. Assemble the peristaltic pump for perfusion following the manufacturer's directions. Using water, confirm that the pump is set for a flow rate of 20 mL/min. Attach the outlet of the pump to a needle for inserting into the rat's heart (see Note 7).
24. Attach the 0.9% NaCl solution to the pump intake and clear all water from the pump lines, making sure to rid the lines of air bubbles. During all subsequent perfusion steps, the pump lines should be kept bubble free.
25. Anesthetize the rats as described in steps 1–3.
26. Place the animal on a perfusion table or near a sink for drainage.
27. With the animal on its back and its tail toward you, use scissors to make a small incision at the base of the animal's abdomen through the underlying muscle.
28. Cut along the left side of the animal toward the head stopping at the rib cage. Repeat for the right side. During the cutting gently pull up to ensure that no organs are damaged.
29. Carefully cut the diaphragm away from the rib cage and then cut along the right and left sides of the rib cage. Next, pull the rib cage away from the chest cavity with a large straight forceps.
30. Clamp the descending artery, which runs adjacent to the spinal cord, with a curved forceps to prevent blood flow to the lower body. This will reduce the volume necessary for effective perfusion of the cranium.

31. Hold the heart with a pair of curved forceps and using the scissors make a small incision in the right atrium of the rat's heart for drainage.
32. With the scissors make a small incision into the base of the heart and insert the needle into the rat's left ventricle toward the left atrium. Hold the needle in place by clamping the curved forceps.
33. Perfuse the rat with 0.9% NaCl at a rate of 20 mL/min until the fluid exiting from the right atrium is clear. Generally, 100 mL of saline is sufficient. Properly and safely dispose of the saline and blood, typically via drainage.
34. Transfer the pump's intake to the ice cold 4% PFA. Perfuse approximately 250 mL total at 20 mL/min. When perfusion begins the animal will twitch. Perfusion is complete when the animal is stiff, especially the head and front legs. Because PFA is toxic be sure to wear gloves when handling and avoid inhaling the fumes. Remove the line from the animal.
35. Transfer the pump's intake back to the 0.9% NaCl solution to flush the PFA from the pump tubing.

Brain Isolation and Dehydration

36. Fill a 20 mL scintillation vial with a 4% PFA.
37. Using a large pair of scissors detach the head.
38. Cut along the midline of the head pulling back the skin with a pair of scissors.
39. Insert the rongeurs forceps into the foramen magnum. Remove skull fragments until the brain is exposed.
40. Gently, using the scissors, remove the meninges overlying the brain, and then with a spatula transfer the brain to the 20 mL vial containing 4% PFA.
41. Incubate brain in 4% PFA for 24 h at 4°C.
42. Dispose of all other tissue as per institutional biosafety requirements.
43. Replace the PFA with a 30% sucrose solution. Initially the brain will be buoyant, and as the sucrose diffuses into the tissue, it will sink to the bottom of the vial. This process can be accelerated by changing the 30% sucrose solution every 12–24 h. Once the brain has sunk proceed to sectioning. During this process store the brain at 4°C, and be sure to minimize the storage time to prevent bacterial growth.

Sectioning

44. Performing the sectioning is dependent on the application and resources available. We generally use a freezing, sliding

microtome and take 40 μm coronal or transverse sections. We prefer to store the sections at −20°C in a 96-well plate format. The plate is filled with 200 μL per well of tissue cryoprotectant (25% ethylene glycol, 25% glycerol and 0.1 M phosphate buffer), and the plate edges are sealed with Parafilm. Under these conditions, the sections can be stored for extended time periods, such as a year.

3.3. Immunostaining

Pretreatment required to expose the BrdU epitope for antibody staining can alter other epitopes, and thereby compromise immunostaining of some proteins (e.g., nestin and GFP). Therefore, if unsure of whether BrdU pretreatment destroys an epitope of interest, it is generally best to first stain for all desired markers followed by BrdU staining. It is helpful to use 12-well net carriers for all rinse steps, since the carriers allow for easy transfer between rinses. However, for antibody incubations transfer the sections to a smaller 24-well plate to help minimize antibody consumption. Transfer and manipulation of sections are easily accomplished using a fine pointed paint brush. Additionally, all steps are performed under gentle shaking, and each net carrier and well in a 24-well plate can hold approximately five sections for the volumes provided.

Antibody Staining

1. Place a net carrier in a 12-well plate and fill wells with 2 mL TBS (pH 7.5).
2. Using a paint brush, transfer desired tissue sections to 12-well plate nets.
3. Shake for 10 min at room temperature and then transfer the section-containing nets to a fresh well of TBS.
4. Repeat 2 mL TBS rinse at room temperature two additional times to remove cryoprotectant. If staining for O4, see Note 8 before proceeding.
5. Block in TBS with 0.3% Triton X-100 and 5% normal serum (TBS-ST) for 2 h at room temperature (see Note 2).
6. Transfer sections to a 24-well plate and incubate with 0.5 mL of primary antibody diluted in TBS-ST for 72 h at 4°C (see Table 1 for suggested dilutions for common markers). Afterwards transfer section back to a 12-well plate with net carriers. Use a paint brush to transfer sections. Thoroughly wash the paint brush with ethanol and rinse between staining sets to prevent antibody cross-contamination. Alternatively, a bent, disposable P200 pipette tip may be used to transfer sections.
7. Rinse twice in 2 mL TBS for 10 min per rinse at room temperature.
8. Rinse once in 2 mL TBS-ST for 10 min at room temperature.

9. Transfer the sections to a 24-well plate and incubate in 0.5 mL of 1:250 diluted secondary antibody in TBS-ST for 1 h at room temperature. While incubating, cover with aluminum foil to protect from light. Care should be taken in all subsequent steps to minimize light exposure and avoid photobleaching. Afterwards, transfer sections back to a 12-well plate with net carriers. Use a paint brush to transfer sections, and use care to prevent antibody cross-contamination.
10. Rinse three times in 2 mL TBS for 10 min per rinse at room temperature.
11. Postfix sections with 2 mL 4% PFA for 1 h at room temperature.
12. Rinse four more times in 2 mL of TBS for 10 min per rinse at room temperature.

BrdU

13. Pretreat for BrdU by incubating sections in 2 mL of 50% formamide solution at 65°C for 2 h using shaker or water bath.
14. Rinse once in 1 mL 1× SSC solution (diluted in deionized water) for 10 min at room temperature.
15. Incubate sections in 2 mL of 2N HCL at 37°C for 30 min using a shaker or water bath.
16. Rinse in 2 mL 0.1 M borate buffer for 10 min at room temperature.
17. Rinse sections six times in TBS (2 mL rinse for a total of 1.5 h).
18. Return to steps 5–10 to perform the primary and secondary antibody staining. The primary antibody is anti-BrdU diluted 1:40 (see Table 1). Using a paint brush, transfer sections to a 24-well plate for antibody incubations to conserve antibody. Thoroughly wash with ethanol and rinse the paint brush between staining sets to prevent antibody cross-contamination. Alternatively, a bent, disposable P200 pipette tip may be used to transfer sections.
19. *Optional*: During the last secondary antibody incubation, DAPI or another nuclear stain, such as TO-PRO®-3, can be used.
20. Mount sections by floating in TBS and transferring with a paint brush to a gelatin-coated slide. Allow the slides to dry in the dark for approximately 30 min (tissue sections will turn translucent when dry).
21. Follow instructions in ProLong® Antifade Kit to prepare antifade/mounting medium. Pipette a line of ~200 μl antifade/mounting medium lengthwise along the edge of the slide,

and slowly lay a coverslip over the slide, starting at the edge with the antifade/mounting medium. Proceed slowly and with care to avoid bubbles. Antifade/mounting medium should evenly coat the slide.

22. Dry the slide at room temperature in the dark for at least 2 h (overnight drying is acceptable). Seal the slide with clear nail varnish. Apply the varnish carefully to avoid covering the samples. Dry at room temperature in the dark for approximately 30 min.

4. Notes

1. It is important to be able to track the cells once they have been engrafted. Depending on the experiment, it is common to use a viral vector (e.g., a retroviral or lentiviral vector) to establish a GFP- or LacZ-expressing cell line. Establishment of this marked cell line via transduction of adult hippocampal neural progenitor cells (AHNPCs) is detailed in Chapter 7. These cells can subsequently be engrafted and tracked without BrdU labeling.
2. Normal serum is serum derived from the species in which the secondary antibody was raised. Use of nonmatched secondary antibodies and blocking serum can lead to nonspecific, background staining.
3. The PBS used during engraftment can be supplemented with growth factors such as FGF-2 or other small molecules as desired. It is best to perform the rinses using the final formulation of PBS used during the engraftment. This ensures that the desired final concentration of supplement is present, since there will always be residual liquid in the cell pellet after each pellet and aspiration step.
4. For sterotaxic coordinates we have used Paxinos and Watson (4). Furthermore, we have found greater accuracy measuring from the dura for the dorsal/ventral positioning of the needle, rather than from the skull surface. For hippocampal engraftments, we have used the following sterotaxic coordinates: AP −3.5, ML ±3.0, and DV −4.0.
5. Cells from Subheading 3.1, step 12 may also be plated on CultureSlides and stained to confirm BrdU incorporation. Use the BrdU staining protocol outlined in Subheading 3.3, steps 13–21 but adjust volumes appropriate to the CultureSlide being used.
6. Within 3–5 min of mixing cells with trypan blue be sure to determine the viability, because longer incubation periods will lead to cell death.

7. A blunted 17- or 18-gauge needle can be used, but we have found that the needle attachment for a hand or bicycle pump works very well.
8. If immunostaining for O4, do not permeablize with Triton X-100. O4 is a surface marker which can be disrupted by Triton X-100. For section staining, follow Subheading 3.3, steps 1–12 without Triton X-100, and then continue with permeablization and costaining at step 5.
9. The mouse anti-O4 antibody is an IgM. Confirm that the intended secondary antibody will recognize IgM before use.

Acknowledgments

We acknowledge Jae Jang and Brandi Ormerod for helpful technical comments. This work was funded by NIH T32 GM007352 (to MJR), an NSF Graduate Research Fellowship (to JP), a California Institute for Regenerative Medicine Training Grant (T1-00007), and NSF BES-0629202.

References

1. Gage FH, Coates PW, Palmer TD, Kuhn HG, Fisher LJ, Suhonen JO, Peterson DA, Suhr ST, Ray J (1995) Survival and differentiation of adult neuronal progenitor cells transplanted to the adult brain. Proc Natl Acad Sci U S A 92:11879–11883
2. Suhonen JO, Peterson DA, Ray J, Gage FH (1996) Differentiation of adult hippocampus-derived progenitors into olfactory neurons in vivo. Nature 383: 624–627
3. Shihabuddin LS, Palmer TD, Gage FH (1999) The search for neural progenitor cells: prospects for the therapy of neurodegenerative disease. Mol Med Today 5:474–480
4. Paxinos G, Watson C (1998) The rat brain in stereotaxic coordinates. Academic Press, San Diego, CA

Chapter 7

Viral Packaging and Transduction of Adult Hippocampal Neural Progenitors

Joseph Peltier and David V. Schaffer

Abstract

Genetic manipulation of adult hippocampal neural progenitor cells is a useful technique for exploring gene function through gain of function and loss of function mutations or RNAi. Furthermore, the introduction of new genes can "re-program" progenitor cell behavior to force a desired lineage in signaling environments that are not normally permissive for that cell fate. Additionally, by using a systems biology approach, neural progenitors can even be taught new behaviors and responses to signaling. In this chapter, we describe protocols for retroviral and lentiviral packaging and transduction of progenitors. These techniques are important for studying the role of various genes in progenitor fate choice.

Key words: Neural, Hippocampus, Progenitors, Rat, *In vitro*, Retrovirus, Lentivirus, Viral transduction, Viral production/purification, Infection

1. Introduction

Genetic manipulation of adult hippocampal neural progenitor cells (AHNPCs) is crucial to understanding the roles various genes play in progenitor fate choice (e.g., self-renewal vs. differentiation). Expressing the constitutively active or dominant negative mutant of an endogenous gene, overexpressing the wild-type gene, or utilizing shRNA expression cassettes to induce RNAi can provide insight into that gene's role in the cellular process of interest. Alternatively, one may wish to transduce cells with reporter constructs (e.g., luciferase or GFP) to readily monitor gene expression.

In this chapter, we describe the methods for packaging retroviral and lentiviral vectors and the subsequent protocols for transduction of AHNPCs *in vitro*. Virus produced in this manner can also be used to transduce AHNPCs *in vivo*. To inject virus into the brain follow the protocol describing AHNPC engraftment in Chapter 6.

Irina M. Conboy et al. (eds.), *Protocols for Adult Stem Cells*, Methods in Molecular Biology, vol. 621,
DOI 10.1007/978-1-60761-063-2_7, © Springer Science+Business Media, LLC 2010

Viral production begins when "helper plasmids" expressing viral structural and enzymatic proteins are cotransfected into human embryonic kidney (HEK) 293T cells along with a "vector plasmid" or "transfer vector" carrying the genetic cargo of interest flanked by two lentiviral or retroviral long terminal repeats (LTR). The deletion of all viral genes between the two LTRs renders the virus replication incompetent. The transfected cells package the vector plasmid cargo into viral particles, which can then be purified and concentrated by ultracentrifugation in a sucrose step gradient. After determining the concentration of infectious units in this concentrate, the virus is ready to transduce cells.

The production and purification processes are identical for both retrovirus and lentivirus; however, the plasmids used for transfection differ. Retroviral production requires structural and enzymatic genes provided by the pCMV *gag-pol* and pcDNA3 IVS VSV-G plasmids (1) in addition to a transfer vector containing the gene of interest flanked by retroviral LTRs. One such transfer vector is pCLPIT (1), which confers puromycin resistance and places the transgene under the control of a tetracycline-controlled transactivator (tTA) resulting in "tet-off" gene regulation (2). Transfer vectors with other selection markers (e.g., GFP, neomycin, or a GFP-puromycin fusion) and/or promoters (e.g., ubiquitin promoter) may be used depending on the intended application.

Lentiviral production requires pMDLg/pRRE encoding the *gag* and *pol* genes (3), pcDNA3 IVS VSV-G (same as retrovirus), and pRSV Rev (3). In addition to the structural genes provided by the above plasmids, lentiviral production also requires a transfer vector containing the gene of interest between lentiviral LTRs, such as pHIV CS (4). Similar to the retroviral vectors discussed above, the lentiviral transfer vector can be modified to express desired selection markers or promoters, depending on the application.

Both retrovirus and lentivirus are capable of efficiently transducing AHNPCs with an insert of up to 9,000 base pairs (5); however, retrovirus will only infect cells that are actively replicating, whereas lentivirus will infect all cells. This makes retrovirus ideal for transducing an AHNPC culture or progenitor cells within a live animal brain, since it will only target the actively dividing progenitors and leave terminally differentiated cells uninfected.

2. Materials

2.1. Viral Packaging

2.1.1. Calcium Phosphate Transfection of 293T Cells

1. 293T HEK cells (ATCC, Manassas, VA). See vendor's instructions for recommended subculturing protocol.
2. 10 cm tissue culture plates.
3. Iscove's Modified Dulbecco's Medium (IMDM, MediaTech, Herndon, VA) with L-glutamine, HEPES, and phenol red;

without α-thioglycerol, β-mercaptoethanol, and sodium biocarbonate.

4. Fetal bovine serum (FBS, Invitrogen, Bethesda, MD). Aliquot and store at −20°C. Avoid freeze/thaw.
5. Penicillin/streptomycin, liquid (PS, Invitrogen). Contains 10,000 U/ml penicillin and 10,000 μg/ml streptomycin. Store at −20°C.
6. Working media (IMDM + FBS/PS). Dissolve 8.84 g IMDM and 1.86 g $NaHCO_3$ in 450 ml deionized water. Sterilize by 0.22 μm bottle-top filtration. Supplement with 50 ml FBS (10% v/v) and 5 ml PS (1% v/v). Store supplemented medium at 4°C for up to 1 month. Warm to 37°C before use.
7. 5 ml round bottom culture tubes (commonly used for flow cytometry) (BD Falcon, Bedford, MA).
8. Sterile tissue culture water.
9. 2.5 M $CaCl_2$ in tissue culture water. Sterilize by 0.22 μm syringe filtration. Store at room temperature.
10. 2× HEPES buffered saline (HeBS). Dissolve 1.64 g NaCl and 1.19 g HEPES in 80 ml tissue culture water. Add 933 μl of 150 mM Na_2HPO_4 solution (tissue culture grade Na_2HPO_4 in tissue culture grade water). pH to 7.05 with NaOH. Sterilize by 0.22 μm syringe filtration. Store at room temperature. Do not refrigerate because refrigeration will adversely affect the pH.
11. For retroviral production, the following high-quality plasmids at 1 mg/ml: pCMV *gag-pol*, pcDNA3 IVS VSV-G, and transfer vector of interest (see Note 1). Plasmids should be purified by Maxiprep purification using a kit from one of many manufacturers (i.e., BioRad, Invitrogen, Qiagen, etc.). For purification instructions, we refer the reader to the manufacturer's Maxiprep kit instructions.
12. For lentiviral production, the following high-quality plasmids at 1 mg/ml: pMDLg/pRRE, pcDNA3 IVS VSV-G, pRSV Rev, and transfer vector of interest (see Note 1). See item 11 for plasmid purification.

2.1.2. Viral Purification by Ultracentrifugation

1. Optima™ L-Series Preparative Ultracentrifuge (Beckman Coulter, Fullerton, CA).
2. SW-41 swinging bucket ultracentrifuge rotor (Beckman Coulter) (see Note 2).
3. Ultracentrifuge tubes (Beckman Coulter). 9/16 × 3½ in Ultra Clear Tubes. Although the manufacturer does not claim sterility, we have found that sterility is maintained if tubes are opened and handled in a biosafety cabinet with good aseptic technique (see Note 3).

4. Working media (IMDM + FBS/PS) (see Subheading 2.1.1, item 6).
5. Flathead screwdriver.
6. Tweezers.
7. 0.45 μm syringe filters.
8. Sterile phosphate buffered saline (PBS), pH 7.4 (Invitrogen). Store at 4°C.
9. 20% w/v sucrose in PBS. Sterilize by 0.22 μm syringe filtration and store at 4°C.

2.2. Quantification of Viral Titer

2.2.1. Titering by Fluorescent Marker Expression

1. 6-well poly-L-ornithine/laminin coated tissue culture plates (see Chapter 5, Subheading 2.1.1 for preparation).
2. AHNPCs (see Chapter 5, Subheading 3.1 for subculturing protocol).
3. Dulbecco's Modified Eagle Medium/Nutrient Mix F-12 (DMEM/F-12) with HEPES and L-Glutamine but WITHOUT phenol red (Invitrogen) supplemented with 1% (v/v) N-2 Supplement (Invitrogen). Store supplemented medium at 4°C for up to 1 month.
4. Accutase (Innovative Cell Technologies, San Diego, CA).
5. Basic fibroblast growth factor (FGF-2, Peprotech, Rocky Hill, NJ). FGF-2 stock solution: Dissolve in DMEM/F-12 + N-2 at 0.1 mg/ml, aliquot, and store at −20°C. Once thawed, aliquot can be stored at 4°C for up to 1 week.
6. Hemacytometer.
7. Sterile PBS, pH 7.4 (Invitrogen). Store at 4°C.
8. 5 ml round bottom culture tubes (commonly used for flow cytometry) (BD Falcon).
9. Flow cytometer (Beckman Coulter FC-500 or equivalent).

2.2.2. Titering by Drug Resistance Selection

1. All reagents in Subheading 2.2.1, items 1–7.
2. 96-well tissue culture plate.
3. Puromycin, cell culture tested (Sigma, St. Louis, MO). Puromycin stock solution: Dissolve in tissue culture water at 1 mg/ml and sterilize by 0.22 μm syringe filtration. Aliquot and store at −20°C. Working aliquots can be stored at 4°C for up to 1 month.
4. Cell Proliferation Reagent WST-1 (Roche, Indianapolis, IN). Upon receipt, thoroughly thaw at 37°C and swirl/pipette to ensure all aggregates dissolve. Then aliquot and store at −20°C. Protect from light. Aliquots can be frozen/thawed if necessary. Before use, warm again to 37°C to dissolve aggregates.

5. Multichannel pipette appropriate for 10 μl volume.
6. 96-well microplate reader.

2.3. Transduction of Adult Neural Progenitors

1. All reagents in Subheading 2.2.1, items 2–6.
2. Poly-L-ornithine/laminin coated culture dishes: 3.5, 6, and 10 cm (see Chapter 5, Subheading 2.1.1 for preparation instructions).
3. Puromycin, cell culture tested (Sigma). See Subheading 2.2.2, item 3 for preparation of puromycin stock solution.

3. Methods

3.1. Viral Packaging

Production of retroviral and lentiviral vectors is accomplished in two steps: (1) Calcium phosphate transfection of 293T cells with all necessary plasmids and (2) concentration and purification by ultracentrifugation through a sucrose step gradient.

Although transfection may be accomplished by other means (e.g., lipofectamine), we have found that the calcium phosphate method gives high transfection efficiencies in 293T cells for very low cost. Approximately 24 h after transfection, virus will begin to be secreted into the culture supernatant, and we recommend harvesting and purifying the supernatant at 48 and 72 h post infection to obtain optimal viral yields.

3.1.1. Calcium Phosphate Transfection of 293T Cells

1. Inoculate 293T cells onto 2 × 10 cm culture dishes in 10 ml IMDM + FBS/PS per plate (see Note 4). Plates should be seeded with approximately 3×10^6 cells/plate (~30% confluent). Incubate overnight at 37°C in 5% CO_2. Cells are ready for transfection once they are attached to the plate, actively dividing, and ~50–70% confluent (see Note 5).
2. For retroviral production: In a 5 ml culture tube, mix 1.72 ml sterile water, 20 μl transfer vector, 12 μl pCMV *gag-pol*, 8 μl pcDNA3 IVS VSV-G, and 2 ml 2× HeBS.

   For lentiviral production: In a 5 ml culture tube, mix 1.72 ml sterile water, 20 μl transfer vector, 10 μl pMDLg/pRRE, 7 μl pcDNA3 IVS VSV-G, 3 μl pRSV Rev, and 2 ml 2× HeBS.
3. Dropwise, add 240 μl of 2.5M $CaCl_2$. Pipette up/down 2–3 times to mix. Incubate at room temperature for 2 min (see Note 6).
4. Dropwise, add 2 ml of the reaction mixture to each 10 cm plate. Incubate at 37°C in 5% CO_2 overnight. Retroviral vectors carrying biohazardous cargo (e.g., oncogenes) and lentiviral constructs in general are considered Biosafety Level 2 biohazards; therefore, proper safety and waste

handling protocols should be in place before proceeding. Consult with your institutional safety committee for approval of all viral vector work prior to initiating studies. All subsequent steps, unless otherwise noted, should be performed at Biosafety Level 2.

5. Replace the media on both plates with 12 ml warm IMDM + FBS/PS 8–14 h post-transfection (see Note 7).
6. Approximately 48 h post-transfection, there is sufficient virus in the supernatant for purification. Proceed to Subheading 3.1.2 for purification by ultracentrifugation (see Note 7).

3.1.2. Purification by Ultracentrifugation

1. Clean and sanitize centrifuge buckets with 70% ethanol. In a biosafety cabinet, spray ethanol into each bucket and aspirate. Then, dry by wrapping lint-free wipes around tweezers, inserting into the bucket, and rotating. Use care to avoid scraping the tweezers on the bucket as they will score the metal. Leave buckets in the biosafety cabinet with the caps off to dry. Ensure that buckets are dry before inserting the centrifuge tubes, because residual ethanol will cause cracking in the centrifuge tubes.
2. Approximately 48 h post-transfection, pool supernatant from the culture plates into a 50 ml centrifuge tube. Replace the media with 12 ml warm IMDM + FBS/PS, and incubate the plates overnight at 37°C in 5% CO_2. This will be used for a second purification the following day.
3. Centrifuge the collected viral supernatant at 1400 RCF for 3 min to remove cellular debris.
4. Filter through a 0.45 μm syringe filter to remove smaller particulates.
5. Pipette 9 ml of filtrate into each of two ultracentrifuge tubes (if using the larger SW-28 rotor, see Note 2).
6. Using 1 ml serological pipettes, add 1 ml of 20% sucrose to the bottom of each tube. Add the sucrose slowly such that a well-defined, clear layer is seen below the viral supernatant. Use care in subsequent steps to prevent disturbing this sucrose layer.
7. Add another 1 ml of filtrate to fill each tube.
8. Seal the tubes in centrifuge buckets and weigh them to ensure that they are properly balanced. Add leftover viral supernatant as necessary to balance (see Note 8).
9. Once the buckets are balanced, re-seal and place them in the rotor. Also seal any unused buckets and place them in the rotor. Place the rotor in the ultracentrifuge, and pull a vacuum in the chamber.

10. Spin at 25,000 rpm (77,000 average RCF) for 1 h 30 min at 4°C (if using the larger SW-28 rotor, see Note 2). To prevent disruption of the sucrose layer during acceleration, set the ultracentrifuge to slowly accelerate to 170 rpm during the first 3 min of the spin. Similarly, set the ultracentrifuge to slowly decelerate from 170 rpm to 0 rpm over 3 min at the end of the spin.
11. Remove the buckets from the ultracentrifuge and bring into a biosafety cabinet.
12. Using tweezers sanitized with 70% ethanol, remove the tubes from the buckets. The virus will be in an invisible pellet at the bottom of the tube.
13. Vacuum aspirate the media layer from each tube by running the aspirating pipette around the wall of the tube in a circular motion as the liquid level is decreasing. Aspirate into the sucrose layer, but do not remove the sucrose layer.
14. Allow residual media to run down the side of the tube into the sucrose layer.
15. Aspirate the remaining liquid with a fresh aspirating pipette by tilting the tube and allowing the liquid to run into the pipette. This prevents accidental aspiration of the viral pellet.
16. Add 100 μl of cold PBS to the centrifuge tube. Keep the virus cold for all subsequent steps. It inactivates relatively rapidly at higher temperatures.
17. Seal with Parafilm® and store at 4°C for at least 2 h to allow the pellet to loosen, but do not incubate longer than 24 h.
18. Repeat step 1 to sanitize the centrifuge buckets.
19. Repeat steps 1–17 the following day to collect more virus.
20. Pipette all viral concentrates up/down several times to ensure that the viral pellet is completely dissociated. Pool viral concentrates in a sterile 1.5 ml tube.
21. Optional: purification for *in vivo* infection (see Note 9). In a fresh ultracentrifuge tube, dilute the concentrated viral stock in PBS to 10 ml total volume. Prepare a balance with 10 ml PBS and follow steps 9–18 and 20 to spin the virus.
22. Aliquot and store at −80°C. Avoid freeze/thaw. Viral stocks are stable at −80°C for approximately 3 months, after which they should be re-titered. Follow Subheading 3.2 to determine viral concentration.

3.2. Quantification of Viral Titer

Accurate knowledge of infectious viral concentration is crucial for a well-controlled transduction. Too much virus will result in multiple infectious events per cell, causing cells to express multiple copies, resulting in poor experimental control and repeatability.

However, insufficient virus will give very few infectious events, potentially making selection extremely difficult.

Depending on the selection marker that is included with the transgene of interest, infectious titer can be determined in one of two ways: fluorescent marker expression or drug resistance. To titer by fluorescent marker expression (e.g., GFP), cells are infected with a dilution series of freshly prepared virus, grown for 2 days, and analyzed by flow cytometry to determine the fraction of fluorescent (infected) cells. Similarly, when titering by drug selection, cells are infected with a dilution series, and then a drug is added (e.g., puromycin) to which infected cells are now resistant. After the uninfected cells have died, the number of remaining cells is quantified and compared to an uninfected control to determine the fraction of infected cells. If using drug resistance, we recommend using puromycin, because it kills all uninfected cells within 3 days, while other drugs can take much longer. For this reason, this protocol is designed for cells expressing puromycin resistance.

It is important to note that the infectious titer is not the same across all cell types, as different cells are differentially susceptible to retroviral and lentiviral transduction. Therefore, we recommend titering virus on the cell type that will be transduced to avoid possible differences in infectivity from one cell type to the next.

3.2.1. Titering by Fluorescent Marker Expression

1. Thaw a 6-well laminin coated culture plate.
2. Prepare AHNPCs. Follow Chapter 5, Subheading 3.1.1, steps 1–11. Determine the cell concentration of the cell suspension by hemacytometer. This solution is generally quite concentrated, so a 1:10 dilution in DMEM/F-12 + N-2 prior to counting is usually necessary. Keep cell suspensions well-mixed throughout the counting and diluting process.
3. Dilute cells to 1.4×10^6 cells in 14 ml (100,000 cell/ml) in DMEM/F12 + N-2. Supplement with 2.8 μl of 0.1 mg/ml FGF-2 stock solution (20 ng/ml final concentration).
4. Aspirate PBS from the culture plate and add 2 ml of cell suspension to each well. Rock the plate several times to mix, but do not swirl. Incubate at 37°C in 5% CO_2 for approximately 4 h.
5. Thaw a viral aliquot in a 37°C water bath for approximately 30 s. If titering retrovirus, we recommend using 16 μl. If titering lentivirus, which is generally produced at a higher concentration, we recommend using 4 μl. Other amounts may be used as necessary.
6. Prepare a 5-point, two-fold dilution series of virus in PBS. For example, if starting with 16 μl of virus, the resulting dilution series will have 16, 8, 4, 2, and 1 μl of virus diluted in PBS such that the final volume of each dilution is 16 μl.

7. Add the dilutions to each well on the 6-well plate. Leave one well as an uninfected control. Gently rock the plate to mix. Incubate for approximately 48 h at 37°C in 5% CO_2.
8. Aspirate media from each well and add 100 μl Accutase. Incubate at 37°C for approximately 2 min. Rock the plate to dislodge the cells.
9. Add 900 μl DMEM/F-2 + N-2 to each well and pipette up/down several times to dissociate cells. Transfer cell suspension into a 5 ml culture tube.
10. Analyze by flow cytometry to determine the fraction of GFP expressing cells (infected).
11. By assuming that infection events follow a Poisson distribution, calculate the multiplicity of infection (MOI, infectious units/cell) at each dilution point (see Note 10). Multiply the MOI by the initial cell number (200,000) to calculate the number of infectious units (IU) in each dilution. Determine viral concentration by calculating the slope of a plot of IU (*y*-axis) vs. volume of virus sample used (*x*-axis) (see Note 11).

3.2.2. Titering by Drug Resistance Selection

1. Follow Subheading 3.2.1, steps 1–5.
2. Follow Chapter 5, Subheading 3.1.1, steps 12–15 to continue culturing leftover AHNPCs. They will be used in 3 days to generate a standard curve.
3. Prepare a 4-point, twofold dilution series of virus in PBS. For example, if starting with 16 μl of virus, the resulting dilution series will have 16, 8, 4, and 2 μl of virus diluted in PBS such that the final volume of each dilution is 16 μl.
4. Add the dilutions to each well on the 6-well plate. Leave two wells as uninfected positive and negative controls. Gently rock the plate to mix. Incubate overnight (12–18 h) at 37°C in 5% CO_2.
5. Add 1.2 μl puromycin stock solution to each well (0.6 μg/ml final concentration) except for one of the uninfected controls. This becomes the positive control. Incubate 48 h at 37°C in 5% CO_2.
6. To generate a standard curve, dissociate, dilute, and count the leftover cells maintained in parallel by following Chapter 5, Subheading 3.1.1, steps 1–11.
7. Prepare standard curve and plate in triplicate 100 μl per well of each cell concentration on a 96-well plate. The cell concentrations can be varied based on typical titers for your specific virus, but we recommend 40,000, 20,000, 10,000, 5,000, 2,500, 1,250, and 0 (DMEM/F-12 + N-2 only) cells per well.

8. Aspirate media from each well of the 6-well plate and add 100 μl Accutase. Incubate at 37°C for approximately 2 min. Rock/shake the plate to dislodge the cells.
9. Add 900 μl DMEM/F-2 + N-2 to each well and pipette up/down several times to dissociate cells.
10. Transfer 100 μl of cell suspension from each well to 3 wells on the 96-well plate. Ensure that the cell suspensions stay well mixed.
11. Thaw WST-1 at 37°C for at least 5 min. Briefly vortex to ensure all aggregates dissolve.
12. Using a multichannel pipette, add 10 μl WST-1 to each well. Gently rock plate to mix. Incubate at 37°C for 2 h.
13. Rock on shaking platform for approximately 1 min.
14. Measure samples against blank (media only) on microplate reader. Measure absorbance at 440 and 700 nm. Subtract A_{700} from A_{440}. This is the experimental readout. For the infected cells, subtract the experimental readout of the uninfected negative control from the readout for the infected wells.
15. Plates can then be returned to the incubator and reanalyzed up to 4 h after WST-1 addition. Use the standard curve to determine the best read. This will depend on the cell concentrations of the standard curve, but we have generally found that 2–3 h post WST-1 addition provides the best results.
16. Use the standard curve to interpolate cell numbers. Divide the number of cells in the experimental wells by the number of cells in the uninfected positive control to determine the fraction of puromycin resistant (infected) cells. Follow Subheading 3.2.1 step 11 to complete the calculations (see Note 11).

3.3. In Vitro Transduction of Adult Neural Progenitors

Here, we describe the transduction and selection of AHNPCs based on puromycin selection. Alternatively, cells can be selected by fluorescence activated cell sorting (FACS) for expression of a fluorescent marker (e.g., GFP); however, we have found FACS to be a harsh procedure for AHNPCs. Approximately 50% of the cells survive the procedure, and those that do survive have long recovery times. This cell death and slow recovery will select for a cell population that is capable of surviving the harsh conditions of FACS. Unfortunately, the characteristics of this population are poorly understood, and furthermore they may not be representative of the typical AHNPC.

1. Thaw a 6-cm and a 3.5-cm laminin coated culture dish.
2. Prepare AHNPCs. Follow Chapter 5, Subheading 3.1.1, steps 1–11. Determine the cell concentration of the cell suspension by hemacytometer. This solution is generally quite

concentrated, so a 1:10 dilution in DMEM/F-12 + N-2 prior to counting is usually necessary. Keep cell suspensions well-mixed throughout the counting and diluting process.

3. Inoculate 750,000 cells into the 6 cm dish in 4 ml of DMEM/F-12 + N-2. Supplement with 0.8 μl of 0.1 mg/ml FGF-2 stock solution (20 ng/ml final concentration).
4. As a negative control, inoculate 200,000 cells into the 3.5 cm dish in 2 ml DMEM/F-12 + N-2. Supplement with 0.4 μl of 0.1 mg/ml FGF-2 stock solution (20 ng/ml final concentration). Dilution of the FGF-2 stock may be required, depending on the accuracy of available pipettes at such a low volume.
5. Incubate 4–8 h at 37°C in 5% CO_2.
6. Add 750,000 IU of virus to the 6 cm plate, resulting in an MOI of 1 IU/cell (see Note 12). Gently rock the plate to mix and incubate 1–2 h at 37°C in 5% CO_2 (see Note 13).
7. Aspirate and replace the medium with 4 ml of fresh, warm DMEM/F-12 + N-2. Supplement with 20 ng/ml FGF-2. Incubate overnight (12–18 h) at 37°C in 5% CO_2.
8. Add 2.4 μl of puromycin stock solution to the 6 cm plate and 1.2 μl to the 3.5 cm plate (final concentration 0.6 μg/ml). Gently rock the plates to mix and incubate at 37°C in 5% CO_2.
9. Depending on the transgene, cellular "behavior" (morphology, proliferation, multipotency, etc.) will vary greatly between viruses. Therefore, monitor the plates daily and take appropriate action as necessary (nothing, replating the cells, passing to a larger/smaller plate, or media change). Some general guidelines:
   (a) Some cell death is normal, depending on the MOI. If the viral titer was correct and infection events follow a Poisson distribution, then at an MOI of 1 IU/cell approximately 37% of cells will remain uninfected and will die upon puromycin selection.
   (b) All cells on the negative control plate should be dead and floating 72 h post puromycin addition.
   (c) Regardless of cell growth/death, replace the media with fresh, warm DMEM/F-12 + N-2 + 20 ng/ml FGF-2 + 0.6 μg/ml puromycin every 48 h.
   (d) If cells are growing in clumps, follow Chapter 5, Subheading 3.1.1 to remove them from the plate, dissociate, and replace on a fresh plate of the same size.
   (e) At very low cell density proliferation will slow significantly. Therefore, if more cell death than expected occurs, it may become necessary to pass the cells down to a

smaller plate (i.e., 3.5 cm dish or even a 12-well plate). Follow Chapter 5, Subheading 3.1.1 for subculturing procedure.

(f) If the cells are >70% confluent, follow Chapter 5, Subheading 3.1.1 to pass all cells to a larger plate.

10. Culture the cells under puromycin for 4 days. At that point, if cell growth has stabilized, they are ready for experimentation. We also recommend cryo-preserving several aliquots. See Chapter 5, Subheading 3.1.2 for cryo-preservation procedure.

4. Notes

1. In addition to encoding the gene of interest, we also recommend including a selection marker into the transfer vector. This marker is crucial for both eliminating uninfected cells and for viral titering. A fluorescent marker, such as GFP, will allow for selection and titering by flow cytometry, cell sorting, or microscopy. However, a drug resistance marker, such as puromycin or neomycin, will allow for selection by cell survival. We have found that a GFP-puromycin fusion protein (6) is an excellent marker, because it combines the ease of titering by flow cytometry with the ease of drug selection.
2. The SW-41 rotor has a working volume of 11 ml (10 ml viral supernatant + 1 ml 20% sucrose). For larger-scale purifications, we recommend the SW-28 rotor, which has a working volume of 35 ml (32 ml viral supernatant + 3 ml 20% sucrose). The centrifugation protocol is similar for the two rotors; however, the SW-28 should be spun at 24,000 rpm, while the SW-41 should be spun at 25,000 rpm.
3. If using the SW-28 rotor, use 1 × 3½ in UltraClear tubes (Beckman Coulter).
4. The number and size of plates may be varied depending on the amount virus required, the application, and the size of the centrifuge rotor available. Each 10 cm plate will yield approximately 10^5–10^8 infectious units (IU); therefore, 2 × 10 cm plates generally provide adequate viral amounts for most *in vitro* applications with a typical MOI of 1 IU/cell. However, most *in vivo* experiments require ~3 μl per animal of viral stock at approximately 10^9 IU/ml; therefore, the number and/or size of plates must be scaled accordingly.
5. Although antibiotics are included in the 293T cultures, good aseptic technique in a laminar flow biosafety cabinet is still required for all viral production and purification steps.

The virus will be used to infect AHNPCs, which are cultured without antibiotics; therefore, great care should be taken to prevent contamination of viral stocks.

6. Although we have generally found that 2 min incubation is ideal, optimal incubation times may vary from 1 to 3 min depending on the batch of HeBS. To check the optimal incubation time, transfect 293T cells with a GFP-expressing vector on a 6-well plate at various time points. Qualitatively assess transfection efficiency the following day by fluorescent microscopy.
7. At 24 h post-transfection cells should look normal. At 48 h post-transfection some cell death and detachment may occur, but the culture should still look mostly healthy. At 72 h post-transfection widespread cell death and detachment may occur. The amount of cell death will depend on the size and type of viral insert.
8. Tubes must be completely full (11 ml for SW-41, 35 ml for SW-28); otherwise, they will collapse during centrifugation. Make up any unused volume with fresh working media.
9. *In vivo* transduction of AHNPCs requires very pure virus to prevent inflammation of the injection site. A second ultracentrifuge spin removes the residual contaminants from the 293T culture media that can cause this inflammation.
10. Assuming a Poisson distribution, the fraction of infected cells is given by

$$x = 1 - e^{-\lambda}$$

where x is the fraction of infected cells, and λ is the MOI (IU/cell). Solving the above equation for the MOI gives

$$\lambda = -\ln(1-x).$$

11. The production/purification protocol written here will typically generate 10^5–10^8 retroviral infectious units for the pCLPIT vector plasmid. Lentiviral amounts are typically 3–4 orders of magnitude higher for a pHIV CS vector plasmid. One of the most significant factors affecting production is the size of the DNA sequence being packaged. Large inserts result in low titers and vice versa.
12. Depending on the application, other MOIs may be used. Note that infectious events follow a Poisson distribution, so at an MOI of 1 IU/cell, approximately 37% of cells will be uninfected, while approximately 26% of cells are multi-infected.
13. Do not leave virus on the cells for more than 2 h. Residual serum in the virus from the 293T culture will induce AHNPC differentiation after prolonged incubation.

Acknowledgment

We acknowledge Matthew Robertson for critical reading of the manuscript. This work was funded by an NSF Graduate Research Fellowship (to JP), a California Institute for Regenerative Medicine Training Grant (T1-00007), and NSF BES-0629202.

References

1. Yu JH, Schaffer DV (2006) Selection of novel vesicular stomatitis virus glycoprotein variants from a Peptide insertion library for enhanced purification of retroviral and lentiviral vectors. J Virol 80:3285–3292
2. Gossen M, Bujard H (1992) Tight control of gene expression in mammalian cells by tetracycline-responsive promoters. Proc Natl Acad Sci U S A 89:5547–5551
3. Dull T, Zufferey R, Kelly M, Mandel RJ, Nguyen M, Trono D, Naldini L (1998) A third-generation lentivirus vector with a conditional packaging system. J Virol 72:8463–8471
4. Miyoshi H, Blomer U, Takahashi M, Gage FH, Verma IM (1998) Development of a self-inactivating lentivirus vector. J Virol 72: 8150–8157
5. Cepko C, Pear W (1996) Overview of the retrovirus transduction system, In: Ausubel FM, Brent R, Kingston RE, Moore DD, Seidman JG, Smith JA, Struhl K (eds) Current protocols in molecular biology, Wiley, New York, Unit 9.9
6. Abbate J, Lacayo JC, Prichard M, Pari G, McVoy MA (2001) Bifunctional protein conferring enhanced green fluorescence and puromycin resistance. Biotechniques 31:336–340

Chapter 8

Culture of Bone Marrow Mesenchymal Stem Cells on Engineered Matrix

Jennifer S. Park, Craig Hashi, and Song Li

Abstract

Mesenchymal stem cells (MSCs) have huge potential to be used in cell therapies because of their pluripotency and immunoregulatory attributes. To harness and maximize the therapeutic potential of MSCs, a thorough understanding of their differentiation pathways and their responses to the microenvironment is needed. The matrix that the cells reside on is an important microenvironmental cue, not only for its chemical components but also for its physical properties. The rigidity and topography of the matrix are two physical factors that are crucial in directing cell behavior. In this chapter, we describe the isolation, culture, and characterization of MSCs. Then we illustrate the procedure of creating different matrix rigidities using acrylamide gels for MSC differentiation study. Finally, we describe how to create MSC-seeded vascular grafts using scaffolds with nano topographical features.

Key words: Mesenchymal stem cell, Bone marrow, Acrylamide gel, Rigidity, Stiffness, Nanofiber, Vascular graft, Topography, Young's modulus

1. Introduction

Mesenchymal stem cells (MSCs) have been shown to differentiate into cells of many tissues such as bone, fat, muscle, tendon, and brain (1–13). Along with their ability to differentiate, MSCs have the ability to inhibit T-cell proliferation and activation (14–16). This immunoregulatory attribute of MSCs has implications in the potential for MSCs as an allogenic source for cell therapies. Although they only make up 0.001–0.01% of cells that reside in the bone marrow (12), MSCs can be expanded a billion-fold in culture. MSCs are most commonly isolated from the bone marrow of the iliac crest, and can also be taken from umbilical cord blood, peripheral blood, placenta, and various other tissues such as adipose tissue (17–20).

Irina M. Conboy et al. (eds.), *Protocols for Adult Stem Cells*, Methods in Molecular Biology, vol. 621,
DOI 10.1007/978-1-60761-063-2_8, © Springer Science+Business Media, LLC 2010

In the 1960s, Friedenstein et al. isolated and identified the MSC population from the bone marrow by discovering their ability to attach to plastic (21). Even today, MSC isolation largely consists of culturing the fraction of bone marrow cells adherent to polystyrene, although density centrifugation using a Percoll gradient is often used to further purify the population (22). These cells are characterized for various surface markers using flow cytometry. Cells can also be tested for their chondrogenic, osteogenic, and adipogenic potential to further confirm their multipotency (12).

Since Friedenstein's MSC study in which he found these cells could differentiate into bone and cartilage in vitro on polystyrene dishes (21), experiments have become more complex in terms of manipulating the matrix that the cells are grown on. Isolating matrix effects in vitro is important to better understand the interactions of the natural ECM with cells in vivo. In order to understand the effect of matrix rigidity, the stiffness of acrylamide gels can be systematically altered by changing the acrylamide or bis acrylamide concentration. MSCs can then be seeded onto these surfaces of varying stiffness. Matrix stiffness can signal the cells along different pathways and by altering signaling, alter proliferation and differentiation.

Another important physical feature of matrix is topography, which controls the morphogenesis of tissues by organizing cells via contact guidance. Aligned nanofibers simulate the matrix fibrils in many tissues, such as blood vessel, heart muscle, skeletal muscle, tendon, and ligament, and cause cells to align along the individual fibers. In addition to morphological control, aligned nanofibers have significant impact on the proliferation and migration of cells. An example is the wall of a blood vessel, where cells in the media layer naturally align in the circumferential direction. Devising vascular grafts out of nanofibrous scaffolds mimics the natural tissue structure and aids in the control of proliferation. In the following protocols, we first describe the isolation, culture, and characterization of MSCs (see Subheading 3.1), followed by protocols detailing the manipulation of matrix rigidity for MSC culture (see Subheading 3.2), as well manipulating matrix topography and constructing MSC-seeded nanofibrous vascular grafts (see Subheading 3.3).

2. Materials

2.1. MSC Isolation, Culture, and Characterization

1. Dulbecco's Modified Eagle's Medium (DMEM, Gibco, Carlsbad, CA). Store at 4°C.
2. Penicillin/Streptomycin 100× (P/S, Gibco, Carlsbad, CA). Store at −20°C.

3. Fetal bovine serum (FBS) (Gibco, Carlsbad, CA). Store at –20°C.
4. Phosphate-buffered saline (PBS), without magnesium and calcium, pH 7.4 (Sigma, St. Louis, MO).
5. Trypsin/0.25% EDTA (Gibco, Carlsbad, CA). Store at –20°C. When ready to use, dilute to 0.5× in PBS and store at 4°C.
6. 0.2% Ethylenediaminetetraacetic acid (EDTA) in PBS (Gibco, Carlsbad, CA).
7. 1% Bovine serum albumin (BSA) in PBS (Sigma, St. Louis, MO). Store at 4°C. To prepare, sonicate powder to dissolve, filter through 0.2-μm filter, heat to 85°C for 10 min, and store solution at 4°C.
8. Human bone marrow (Lonza, Basel, Switzerland (formerly Cambrex)) (see Note 1). *Important*: If bone marrow is obtained from human subjects, a statement confirming that informed consent must be obtained from all subjects.
9. Human mesenchymal stem cells (Lonza, Basel, Switzerland (formerly Cambrex)) (see Note 1).
10. Mesenchymal stem cell growth medium (MSCGM, Lonza, Basel, Switzerland (formerly Cambrex)) (see Note 2).
11. Percoll Solution (Sigma, St. Louis, MO) (see Note 3).
12. 1.5 M NaCl.
13. 4% Acetic acid. CAUTION: Corrosive. Prepared in a chemical hood. Wear protective goggles, clothing, and gloves as appropriate.
14. Antibodies against CD14, CD29, CD34, CD44, CD45, CD105, CD166 (see Note 4).
15. Isotype negative control.
16. Fluorescent-labeled secondary antibody.
17. 4% Paraformaldehyde in PBS (Fisher Scientific, Pittsburgh, PA). To prepare, heat the buffer to 55–60°C to dissolve powder, filter through 0.2-μm filter, and store at 4°C. Before filtering, NaOH can be added to help dissolve powder. CAUTION: Irritant and carcinogen. Wear protective goggles, clothing, and gloves as appropriate.
18. 0.5% Triton-×-100 in PBS (Sigma, St. Louis, MO).
19. Flow cytometer.
20. Round-bottomed tubes for flow cytometry.
21. 15-ml centrifuge tubes.
22. 100-mm round cell culture dishes.
23. Pasteur pipettes.

2.2. Engineering the Mechanical Property of ECM for MSC Culture

2.2.1. Silane Slide Coating (see Note 5)

1. (3-Aminopropyl)tris(trimethylsiloxy)silane (Sigma, St. Louis, MO). Avoid contact with air. CAUTION: Skin and eye irritant.
2. 0.1 N NaOH.
3. 70% ethanol. CAUTION: Flammable. Stored in a flammable-safe cabinet.
4. Plain glass slides.
5. Plastic slide holders (for histology; holds 24 slides).
6. Plastic slide holder containers (for histology; holds liquid for dipping slides in slide holder).
7. Square dishes, large, 234 mm (Fisher Scientific, Pittsburgh, PA).
8. Parafilm.
9. Sonicator.

2.2.2. Acrylamide Gel Polymerization

1. 0.5% Glutaraldehyde. CAUTION: Toxic irritant. Prepare in fume hood and wear appropriate skin and eye protection.
2. Gel Slick Solution (Lonza, Basel, Switzerland (formerly Cambrex)).
3. 30% Acrylamide (Polysciences, Inc, Warrington, PA).
4. 2% Bis-acrylamide (Polysciences, Inc, Warrington, PA).
5. 10% Ammonium persulfate (APS) (Bio-Rad, Hercules, CA).
6. N,N,N′,N′-Tetramethylethylenediamine (TEMED) (Sigma, St. Louis, MO).
7. 1 M HEPES, pH 7.
8. Plastic coverslips (Chemicon, Temecula, CA).
9. Glass coverslips 24×60 mm (VWR, San Diego, CA).
10. Square dishes, large, 234 mm (Fisher Scientific, Pittsburgh, PA).
11. Kimwipes.
12. Vacuum dessicator.

2.2.3. ECM Conjugation to Acrylamide Gel

1. 50 mM HEPES, pH 7.
2. 50 mM HEPES, pH 8.5 (see Note 6).
3. Sulfo-SANPAH (*N*-Sulfosuccinimidyl-6-[4′-azido-2′-nitrophenylamino] hexanoate) (Pierce Biotechnology, Rockford, IL). Store at −20°C. CAUTION: Avoid exposure to light. Prepare fresh in small volume of DMSO to help dissolve (0.1 mg/μL). Once dissolved, dilute with 50 mM HEPES, pH 7.
4. Orbital shaker.
5. UV table (320–350 nm wavelength).
6. Collagen I (BD Biosciences, San Jose, CA) (see Note 7).
7. Phosphate buffered saline (PBS), without magnesium and calcium, pH 7.4 (Sigma, St. Louis, MO).

2.2.4. Cell Seeding on Acrylamide Gels

1. Cell culture reagents (see Subheading 2.1): DMEM, Penicillin/Streptomycin, FBS, PBS, Trypsin/0.25% EDTA.
2. Cell culture materials (see Subheading 2.1): 15-ml and 50-ml centrifuge tubes, 100-mm round cell culture dishes, Pasteur pipettes.
3. Square dishes, medium, 100 mm (Fisher Scientific, Pittsburgh, PA) (see Note 8).

2.2.5. Measuring Young's Modulus of Acrylamide Gels

1. Comparator (B.C. Ames Company, Waltham, MA)
2. 50 g cylindrical weight (different weights can be used depending on gel stiffness, see Note 9).
3. Round plastic disk, 3 cm in diameter.
4. 6-well plate.
5. Materials for making gels (see Subheading 2.2.2): 30% acrylamide, 2% bis acrylamide, 10% APS, TEMED, 1M HEPES, pH 7.

2.3. MSC-Seeded Vascular Grafts for Tissue Engineering

1. Cell culture reagents (see Subheading 2.1): DMEM, penicillin/streptomycin (P/S), FBS, PBS, Trypsin/0.25% EDTA.
2. Nanofiber sheets ~100–250 μm in thickness.
3. Fibronectin, bovine-derived (Sigma, St. Louis, MO). Prepare 0.1% weight/volume) in PBS.
4. 70% Ethanol. CAUTION: Flammable. Store in a flammable-safe cabinet.
5. Forceps.
6. 7-0 nylon sutures.
7. 15-ml tube.
8. Stainless-steel mandrel.

3. Methods

3.1. MSC Isolation, Culture, and Characterization

Important: MSC isolation and culture steps should be performed in a laminar flow hood using sterile technique and sterile reagents.

3.1.1. Isolation of MSCs from Bone Marrow by Adherence Only (see Note 10)

Filter bone marrow aspirate through a 70-μm nylon mesh.

Plate cells on round 100-mm tissue culture dishes at a density of $1.7 \times 10^6/\text{cm}^2$.

After 24 h, remove non-adherent cells by aspirating away the media.

Add fresh MSCGM to adherent cells in dish. All non-adherent cells may not be removed after the first feeding. However, within several days of subsequent feedings, the non-adherent cells will be removed completely.

3.1.2. Isolation of MSCs from Bone Marrow by Density Centrifugation (see Note 10)

Add media to marrow aspirate 1:1 in a 50-ml centrifuge tube. DMEM containing 10% FBS and 1% P/S can be used for isolation steps as opposed to the more expensive MSCGM since the media in these steps is used mainly for washes at short time intervals.

To count the cells, transfer 200 μl of the marrow aspirate to a microcentrifuge tube. Dilute the mixture 1:1 (v/v) with media. Add 4% acetic acid at a 1:1 ratio (v/v) to the diluted mixture in order to lyse erythrocytes. Mix well before counting the nucleated cells with a hemacytometer.

Centrifuge the marrow solution at 500×*g* for 5 min.

Aspirate fat and supernatant, being careful not to disturb the cell pellet.

Prepare 35 ml of Percoll gradient solution: 22.05 ml of Percoll, 2.45 ml of 1.5 M NaCl, and 10.5 ml of PBS (see Note 3).

Spin Percoll solution at 21,200×*g* for 15 min in a fixed-angle centrifuge (see Note 11). The resulting density gradient is equal to 1.03–1.12 g/l.

Place 2×10^8 cells in 5–35 ml of Percoll solution (very carefully at the top of the solution). If the cell solution exceeds 5 ml, put the excess in another tube of 35 ml Percoll solution.

Spin in fixed-angle centrifuge at 400×*g* for 15 min with the brake off (see Note 11).

Collect the nucleated cells at the top (10–14 ml) in a fresh centrifuge tube and add media to 50 ml. Discard red blood cells at the bottom of the tube.

Spin new tube at 500×*g* for 5 min in centrifuge.

Remove supernatant and resuspend cells in 10 ml media.

Plate cells at $1.7 \times 10^6/\text{cm}^2$. Non-adherent cells will be removed with subsequent feedings. Adherent cells will remain attached to the dish.

After about three feedings with MSCGM when the adherent population is at least 90% pure, feed cells with MSCGM every 2–3 days (see Note 2).

Seed cells at 3,000–4,000 cells/cm^2 and grow to 80% confluency (10,000 cells/cm^2) before sub-culturing.

3.1.2.1. Subculturing Cells

1. Aspirate media from the cell culture dish.
2. Wash with PBS.
3. Add 3 ml of 0.5% trypsin/EDTA to 100-mm round dish.
4. Incubate at 37°C for 3 min or longer until about 90% of the cells detach.
5. Tap plate to remove remaining cells being careful not to spill liquid.
6. Add 3 ml of MSCGM to deactivate trypsin activity (see Note 2).

7. Collect cells and centrifuge at 500×*g* for 5 min.
8. Aspirate the supernatant, being careful not to disturb cell pellet.
9. Resuspend cell pellet in appropriate amount of MSCGM and break up cell pellet by pipetting up and down.
10. Plate cells onto three dishes. Cells are generally subcultured at a dilution ratio of 1: 3 to give a density of 3,000–4,000 cells/cm^2.

3.1.3. Flow Cytometry to Characterize MSCs After Isolation and/or Passaging

Culture MSCs in two 100-mm round cell culture dishes to 80% confluency. Each plate can be divided into 3–4 samples for FACS analysis (roughly 10^5 cells per sample) (see Note 12).

Aspirate media from cell culture plates and wash with PBS.

Detach cells using 3 ml of 0.2% EDTA. (Avoid using trypsin, which can remove cell surface proteins).

Incubate at 37°C for 5 min.

Add 3 ml DMEM with 10% FBS to each dish and collect cells in centrifuge tube.

Centrifuge at 500×*g* for 5 min.

Aspirate supernatant and resuspend cells in 5 ml PBS, breaking up cell pellet.

Centrifuge at 500×*g* for 5 min and resuspend cells in 1 ml 1% BSA to block nonspecific binding.

Incubate on ice for 30 min.

During the BSA incubation step prepare round-bottom test tubes containing appropriate antibody volumes for CD14, CD34, CD45, CD44, CD29, CD105, CD166, or 100 μl isotype negative control (see Note 4 and 13). Add PBS to obtain a total of 100 μl volume in each tube.

Centrifuge the 50-ml tube containing cells at 500×*g* for 5 min and resuspend cells in 5 ml PBS.

Centrifuge at 500×*g* for 5 min and resuspend cells in 800 μl PBS. Pipette 100 μl into each of the prepared tubes from step 10.

Flick tubes to mix and incubate on ice for 30 min.

Add 2 ml PBS to each of the eight tubes and centrifuge at 500×*g* for 5 min.

Aspirate supernatant and resuspend cells in 5 ml PBS. *Important*: When aspirating liquid, be careful not to disturb the thin layer of cells on the bottom of round-bottom test tubes.

Centrifuge at 500×*g* for 5 min and resuspend each tube in 200 μl fluorescent-conjugated secondary antibody diluted 1:50 in PBS.

Protect tubes from light and incubate on ice for 30 min.

Repeat steps 14–15 to wash cells in PBS.

Centrifuge at 500×*g* for 5 min and resuspend cells in 0.5 ml PBS, thoroughly breaking up cell clumps. *Important:* Pipette cells repeatedly, about ten times, in order to dissociate cells by mechanical stress and obtain a single cell suspension for FACS analysis. 10,000 single cell events should be counted for an adequate cell population reading.

3.2. Engineering the Mechanical Property of ECM for MSC Culture

Figure 1 gives a graphical representation of the stages of performing these steps. Figure 1a represents silane slide coating (see Subheading 3.2.1), Fig. 1b represents acrylamide gel polymerization (see Subheading 3.2.2), Fig. 1c represents ECM conjugation to acrylamide gel (see Subheading 3.2.3), and Fig. 1d represents cell seeding on acrylamide gels (see Subheading 3.2.4). Figure 2 gives an overall timeline of the day of the experiment starting from acrylamide gel polymerization (see Subheading 3.2.2) to cell seeding on acrylamide gels (see Subheading 3.2.4).

3.2.1. Silane Slide Coating (see Note 5 and Fig. 1a)

Place 24 glass slides in plastic slide holder and dip in plastic container filled with distilled water (ddH_2O), then into plastic container with 70% ethanol.

Sonicate slides in ethanol for 15 min.

Air-dry slides in slide holder.

While slides are sonicating/drying, clean two large square dishes with ethanol.

Lay down parafilm (cleaned with ethanol) in two columns covering the length of each dish. Smooth out any air bubbles under parafilm with a kimwipe (see Note 14).

Pipette 150 μl of 0.1 N NaOH on parafilm in six spots with even spacing down each column (12 spots total).

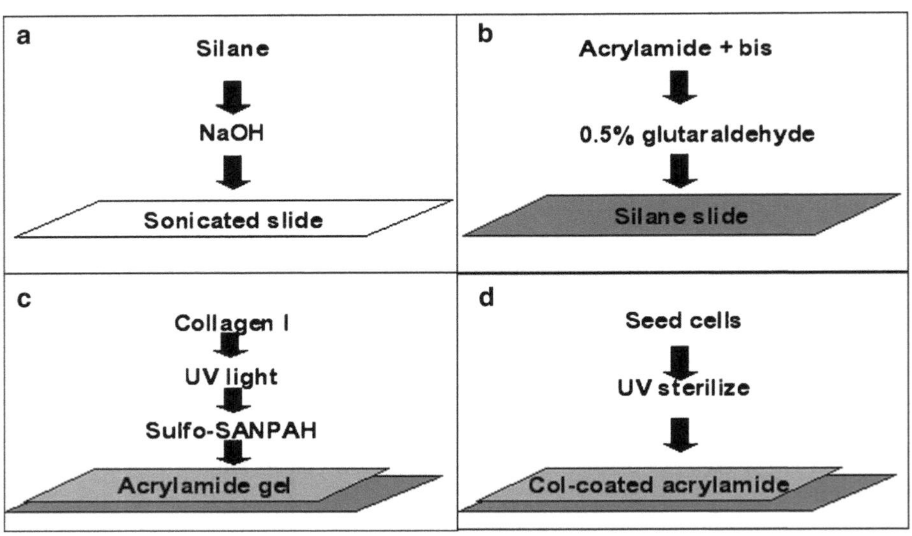

Fig. 1. Procedure to manipulate matrix rigidity and seed cells. Slides are silanated (**a**) so that glutaraldehyde can link acrylamide gels to the slide (**b**) Sulfo-SANPAH is used to cross-link collagen to the acrylamide gel (**c**) and cells are seeded onto the collagen-coated acrylamide gels (**d**) a can be done up to 2 weeks ahead of time. **b–d** can be done in 1 day or broken up into multiple successive days

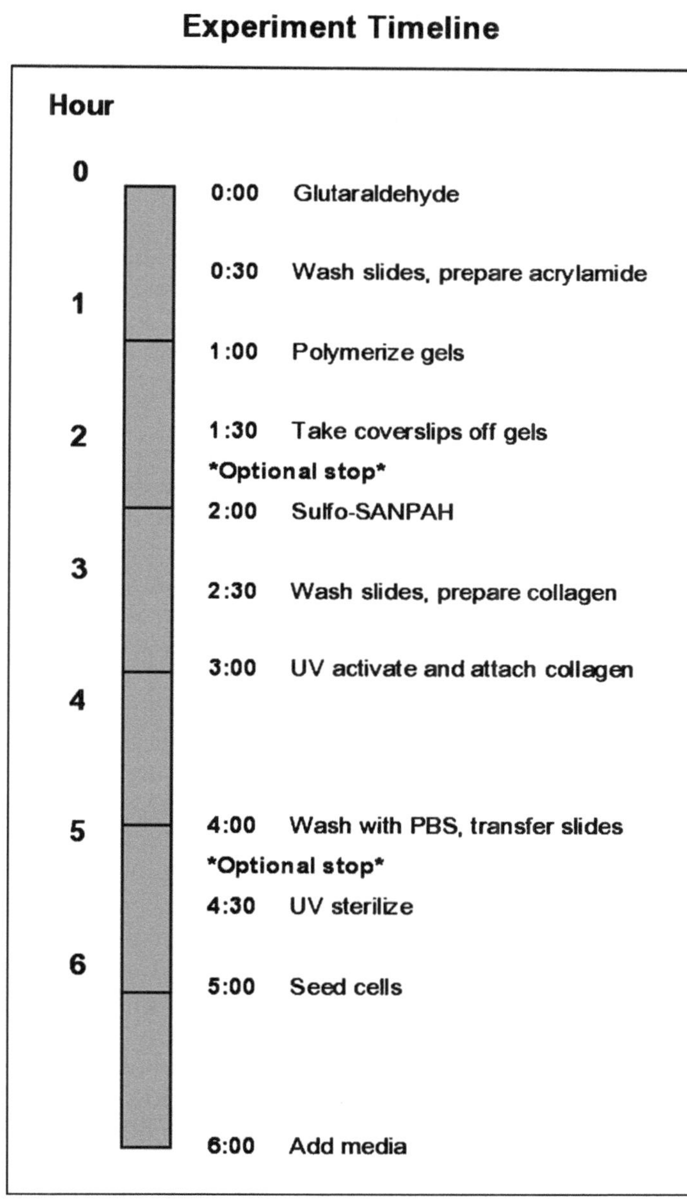

Fig. 2. Timeline for making gels, attaching ECM and seeding cells onto acrylamide gels. Timeline starts after silaned slides are made (see Subheading 3.2.2). It is recommended to perform the complete time line in 1 day; however, optional stopping points exist

Place glass slides on each pool of NaOH avoiding air bubbles getting trapped underneath and let sit for 5 min (12 slides per tray).

Store slides in holder with NaOH-treated side facing middle of the rack to keep track of coated side.

Dab the bottom of the slides with a kimwipe to dry excess fluid. Slides need not be completely dry.

Pipette 150 μl of silane solution onto parafilm on other dish in the same manner as NaOH. Act quickly since silane is sensitive to air. Close cap of silane bottle tightly. Carefully lay down the slides as described in Step 7 and let sit for 3 min.

After 3 min, wash silane off with ddH_2O. Cover slides with ddH_2O in tray and rub the back of each slide thoroughly in stream of water (see Note 15). Rinse the top of slide gently making sure that there is no hardened silane residue remaining.

Place slides in holder with treated side facing middle and air-dry in hood.

Once dry, wrap in foil until day of experiment. Slides can be stored for a few weeks at room temperature.

3.2.2. Acrylamide Gel Polymerization (Fig. 1b)

Clean a large square dish with ethanol. Place silane-treated slides with silane-side up in the dish.

In fume hood pipette 300 μl of 0.5% glutaraldehyde on one end of each slide. Place ethanol-cleaned plastic coverslips on each slide to spread glutaraldehyde, leaving approximately 1.5 cm of unexposed glass on one side and being careful not to spill the liquid underneath the slide (see Note 16).

Wait for 30 min.

Meanwhile, determine total amount of acrylamide gel solution required (0.4 ml of acrylamide solution per slide for ~250 μm thickness gel), and measure out desired ratios of acrylamide and bis-acrylamide, adding 1M HEPES (pH 7) and ddH_2O so that solution has a final concentration of 10 mM HEPES (see Note 17).

Place tubes of acrylamide/bis-acylamide solution in vacuum dessicator for 15 min.

After 30 min of glutaraldehyde treatment (step 3), wash the slides with ddH_20, making sure to remember which side is unexposed to glutaraldehyde (see Note 18). Wash the slides a second time with ddH_2O.

Space slides evenly on dish and let dry in a ventilation hood. Aspirate water near the edges of the slides.

Meanwhile, rub 24×60 mm glass coverslips with Gel Slick solution and wipe down with ethanol to remove dust/streaks. Prepare the same number of coverslips as slides (see Note 19).

If there is still water left on the slides, aspirate the rest.

Take the acrylamide solutions out of the vacuum dessicator.

To most dilute solution (softest rigidity), add APS (1:167) and TEMED (1:250) to gel solution, mix well (avoiding bubbles), and pipette 0.4 ml of solution on each slide. Act quickly to avoid polymerization before pipetting.

Immediately cover each slide with a Gel Slick-treated glass coverslip (Gel Slick side down), laying down one end of the coverslip first (see Note 20) and gently dropping the rest of the coverslip on the slide, making sure that liquid covers slide evenly without any air bubbles (see Note 21).

Repeat for all solutions, working faster as rigidity increases since the polymerization is faster.

Wait 30 min (or until you can see clear outlines of the slightly opaque 6% acrylamide, 0.5% bis acrylamide gel) and pour 50 ml of 200 mM HEPES (pH 8.5) on dish to cover slides.

Take off glass coverslips, using forceps or fingernail to lift up corner, slowly lifting up the rest of the coverslip (see Note 22).

Proceed to Subheading 3.2.3. *Optional*: If necessary, gels can be stored overnight in HEPES buffer at 4°C and the protocol can be continued in the next 1–2 days.

3.2.3. ECM Conjugation to Acrylamide Gel (Fig. 1c)

Measure out 1 mg/slide of Sulfo-SANPAH in plastic weigh boat (see Note 23). *Important*: Sulfo-SANPAH is moisture- and light-sensitive. Let it come to room temperature first before opening bottle. Minimize exposure to light.

Pipette 10 μl of DMSO per mg of Sulfo-SANPAH in weigh boat in order to dissolve powder. Sulfo-SANPAH is insoluble in HEPES by itself.

Add 200 mM HEPES (pH 8.5) to make enough volume for 0.5 ml/slide. Sulfo-SANPAH attaches best at neutral pH. *Important*: Wrap the tube in foil to protect from light.

Pour out excess liquid in dish and space slides apart with forceps so that they are not touching each other (see Note 24).

Pipette 0.5 ml of Sulfo-SANPAH solution onto each slide, making sure that the solution is spread on the entire area of the gel but does not fall off the slide.

Cover dishes with foil and incubate for 30 min.

Prepare 200 mM HEPES (pH 8.5) (see Note 6). About 500 ml is required to wash 24 slides.

After 30 min, pour out Sulfo-SANPAH residue and pour ~100 ml of 200 mM HEPES (pH 8.5) in each dish to cover slides.

Shake the dishes on orbital shaker for 5–10 min (covered in foil) and change the washing buffer until the red-orange color diffuses out of gels (two to three times is sufficient).

Meanwhile, prepare plain glass slides for controls. Sonicate desired number of glass slides in plastic slide holders in ethanol.

Dilute collagen I in 0.1% acetic acid at 0.2 mg/ml. Make enough solution for 0.5 ml/slide. (100 μg/slide) (see Note 7)

After Sulfo-SANPAH is washed off, pour out liquid and put uncovered dish underneath an upside-down UV table (so that UV light is shining down, 3–4 in. from dish). Leave the dish under UV for 5 min.

After 5 min, quickly transfer photoactivated slides in dish to the tissue culture (TC) hood, spacing apart slides with foceps and aspirating excess liquid (being careful to avoid aspirating gels as stated in Note 24).

Immediately pipette 0.5 ml of diluted collagen solution (from step 11) on each slide (see Note 25).

Repeat with control glass slides. Control glass slides can be placed in 100 mm×100 mm square dishes, three slides in a dish, where they will eventually be cultured. Leave 1.5 cm space at the end of each slide uncoated with collagen to achieve the same area as slides with the gels with the same uncoated area. Mark the bottom of the dish to distinguish the slides with different coatings. Do not mark the slides directly (see Note 18).

Leave collagen on for 30 min to 1 h in TC hood (no UV light) (see Note 26).

Rinse slides with gels in PBS and transfer the slides to 100 mm×100 mm square dishes, three slides per dish. Rinse control glass slides.

Leave the slides in PBS in the TC hood under UV for 15 min to sterilize.

Proceed to Subheading 3.2.4. *Optional*: If necessary, gels can be stored overnight in PBS and protocol can be continued the next day (see Note 27).

3.2.4. Cell Seeding on Acrylamide Gels (Fig. 1d and 3)

Trypsinize cells (refer to Subheading 3.1.2, steps 1–8) using one round 100-mm dish per 12 slides to achieve ~70% confluency (see Note 28).

While spinning cells down, aspirate PBS off gels and space apart. *Important*: Do not do this too early or the gels will dry out, making cell solution difficult to spread. Start with control glass slides and go from most rigid to least rigid gel, with the softest gels aspirated right before cell solution is ready as they dry out the quickest. For control glass slides, make sure to dry the same 1.5 cm space on the end of the slide uncoated with collagen (see Note 29).

Resuspend cells in DMEM containing 10% FBS and 1% P/S (see Note 30).

Pipette 0.5 ml on each slide. First check confluency on one slide under the microscope to make sure seeding density is correct. Be careful to spread cell solution all over gel, but not to let the solution fall off the slide. *Important*: It is critical that the slides be dried correctly so that the cell solution remains on the slide. Once the solution falls off, it is extremely difficult to put the cells back on and there may be cells that cannot be retrieved from the dish, making seeding uneven.

Transfer slides to incubator and let attach for 30 min–1 h.

Add media so that each square dish holds 10 ml media (see Note 8). Cell morphology should look different on different rigidities (Fig. 3).

3.2.4.1. Measuring Young's Modulus of Acrylamide Gels

Young's modulus is a quantitative measurement of stiffness, and can be measured by various methods and devices such as an atomic force microscope or an Instron machine, as long as the deformation of the object can be measured accurately in response

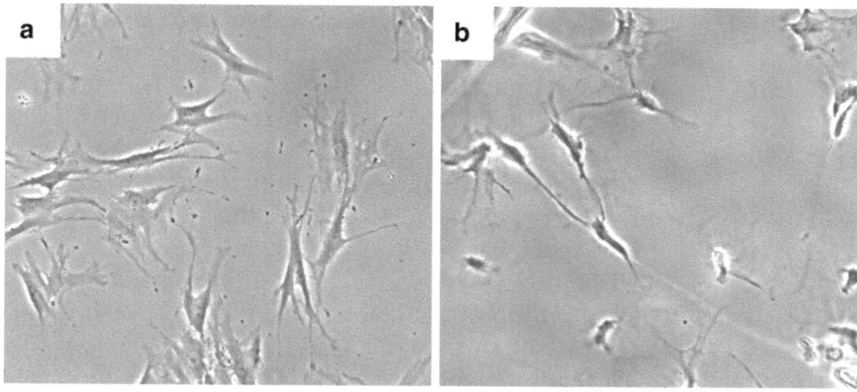

Fig. 3. Phase contrast microscopy images of MSCs seeded on glass (**a**) and acrylamide gel (**b**)

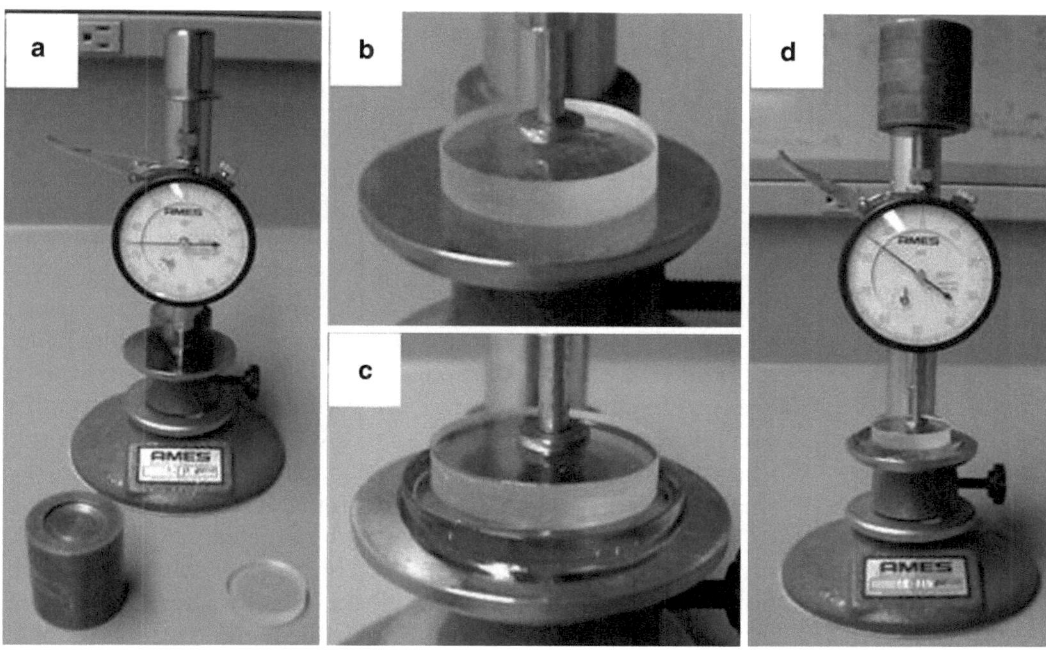

Fig. 4. Measuring Young's modulus of acrylamide gel using a comparator. This procedure requires a comparator, weight, and disk (**a**) First the disk is placed on the stage and the distance measurement is read (**b**). Then the gel is placed underneath the disk and the measurement is read (**c**). Finally the weight is placed on top of the comparator which compresses the gel and the measurement is read. The displacement is determined by subtracting the readings between (**c**) and (**d**) and the initial height of the gel is determined by subtracting the readings between (**b**) and (**c**). With these displacement values, the area of the disk and the known force of the weight, the Young's modulus can be calculated

to a known force over a known area. We describe how to measure Young's modulus using a comparator, a simple device with a dial that reads the initial position of the object of interest and the subsequent position (displacement) given an applied force on that object (Fig. 4a). A disk is placed on the object to provide a flat non-deformable object of known area that distributes the weight evenly across the object that you are measuring, in this case an acrylamide gel.

Make enough acrylamide gel, about 3 ml per 6-well for each gel rigidity (see Note 31).

Place the plastic disk on the comparator stage and take the reading on the dial (Fig. 4b). This is the initial distance (d_0). Make sure the metal bar is resting on the disk.

Place the gel on the stage of the comparator with the disk centered on top of the gel, and the metal bar resting on the disk (Fig. 4c). Take the reading, (d_1).

Place the weight on top of the comparator and take the final displacement reading (d_2) (Fig. 4d, see Note 9).

Calculate the Young's modulus, E = stress/strain

$$E = \frac{\text{Force} / \text{Area}}{\Delta\text{length} / \text{original length}}$$
$$= \frac{\text{mass of weight}(\text{kg}) \times 9.8 \text{ m} / \text{s}^2 / \text{area of disk}}{(d_2 - d_1) / (d_1 - d_0)}$$

3.3. MSC-Seeded Vascular Grafts for Tissue Engineering

The nanofibers used in this protocol can be created in any standard laboratory with the aid of a high voltage generator and syringe pump through a process known as electrospinning (23). Electrospinning works by creating an electric field between a charged polymer solution and a collection plate, which are spaced anywhere from two to thirty centimeters apart. The polymer solution (the concentration and specific type of polymer can be changed to suit desired needs, see Note 32) is ejected through a small diameter needle connected to a syringe pump. The positive lead of the high voltage generator is electrically coupled to the small diameter needle. The collection plate is electrically coupled to a grounded source, thereby creating an electric field between the needle and the collecting plate. The polymer solution is drawn by the electric field, resulting in fibers with diameter ranged from tens of nanometers to several microns, depending on the voltage, distance between the collection plate and the needle, the concentration of the polymer, and the flow rate of the polymer solution.

Using this protocol, the nanofibers will be randomly oriented. The fibers can be aligned during the spinning process or post-fabrication (Fig. 5a). Alignment during the spinning process involves a more complex system, and therefore will not be discussed in this protocol. Post-fabrication alignment can be easily achieved by straining the nanofiber sheets in a particular direction, while submerged in a water bath set above the glass transition temperature of the polymer. The alignment of nanofibers can be verified in a number of ways, including phase contrast microscopy, scanning electron microscopy, and reflection microscopy. Cell alignment on nanofibers can also be used as

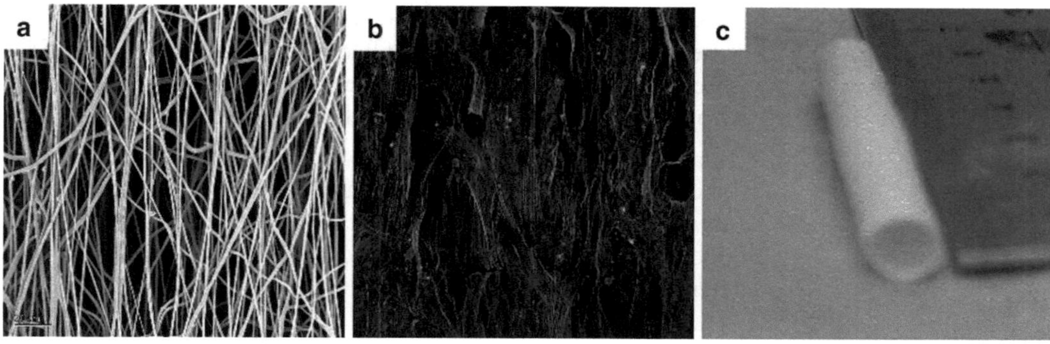

Fig. 5. (**a**) Scanning electron micrograph of vertically aligned nanofibers. (**b**) MSCs seeded on vertically aligned nanofibers. MSCs were stained for F-actin with FITC-conjugated phalloidin. (**c**) Electrospun tubes with nanofibrous structure

indirect demonstration of nanofiber alignment. Nanofibrous sheets can be used as scaffolds for cell seeding and tissue construction.

Important: All steps should be performed in a laminar flow hood using sterile technique and sterile reagents.

3.3.1. Fibronectin Coating and MSC Seeding

1. Sterilize nanofiber sheets by soaking them in 70% ethanol while under UV light in TC hood for 30 min.
2. Wash sheets with autoclaved ddH_2O three times. *Important*: Do not let the nanofiber sheets dry out. Keep moist with PBS or ddH_2O.
3. Wash sheets once with sterile PBS.
4. Incubate sheets with 5 μg/cm^2 fibronectin at 37°C for 30 min.
5. Wash sheets once with PBS.
6. Count cells using hemacytometer.
7. Seed MSCs at ~100,000 cells/cm^2 onto the nanofiber sheet using 300 μl of cell suspension per cm^2 of nanofiber sheet (similar cell seeding can be performed using alternative cell types, see Note 33).
8. Incubate samples for 2 days in DMEM with 10% FBS and 1% P/S.

Optional: The morphological characteristics and cell seeding density can be observed at this stage by fixing the sample and staining with various fluorescent stains, such as FITC-phalloidin (Fig. 5b). It is not recommended to use hematoxylin and eosin or any other immunohistochemistry method.

3.3.2. Vascular Graft Construction

1. Sterilize latex gloves, forceps, and stainless steel mandrel by autoclaving.
2. Wear sterilized latex gloves over ethanol-sprayed latex gloves.

3. Use forceps to transfer MSC-seeded nanofiber sheets from the media into a new dish filled with PBS, with the MSC-seeded side of the nanofiber sheet facing up. *Important*: Do not let samples dry out, wet with PBS when necessary.
4. Place the stainless steel mandrel at one edge of the nanofiber sheet, such that the mandrel is parallel to the edge of the sheet. *Important*: Be delicate when handling the nanofiber sheet. Do not press hard when placing the mandrel onto the nanofiber sheet.
5. Gently turn the mandrel while rolling the nanofiber sheet around the mandrel, such that the MSC-seeded side will be on the luminal side of the graft.
6. Continue rolling until the entire sheet is around the mandrel.
7. Use sterile 7-0 nylon sutures to bind the free end of the nanofiber sheet by tying the sutures around both the mandrel and nanofibers, spaced every 5 mm apart.
8. Carefully remove the nanofiber tubular construct by sliding it off the mandrel while keeping the sutures intact.
9. Place the construct into a 15-ml tube filled with 5 ml of DMEM with 10% FBS and 1% P/S.
10. Incubate for 2 days at 37°C with the 15-ml tube loosely capped thereby allowing carbon dioxide gas exchange to the media and cells.
11. Just prior to implantation, wash the construct with PBS once.
12. While the construct is being anastomosed to the artery or vein, periodically irrigate the cavity and construct with sterile saline solution.

3.3.3. Vascular Graft Construction (Alternative Method)

1. Instead of using a nanofiber sheet, electrospin nanofibers directly onto a mandrel, thereby creating a tubular construct (Fig. 5c).
2. Remove the tubular construct from the mandrel.
3. Sterilize nanofiber construct by soaking it in 70% ethanol while under UV light in TC hood for 30 min.
4. Wash construct with autoclaved ddH_2O three times. *Important*: Do not let the nanofiber construct dry out. Keep moist with PBS or ddH_2O.
5. Wash once with sterile PBS.
6. Incubate construct with 5 μg/cm^2 fibronectin at 37°C for 30 min.
7. Wash construct once with PBS.
8. Count cells using hemacytometer.

9. Seed MSCs at ~100,000 cells/cm^2 onto the nanofiber construct in a rotating microcentrifuge tybe.
10. Place the construct into a 15-ml tube filled with 5 ml of DMEM with 10% FBS and 1% P/S.
11. Incubate for 2 days at 37°C with the 15-ml tube loosely capped, thereby allowing carbon dioxide gas exchange to the media and cells.

    Optional: The morphological characteristics and cell seeding density can be observed at this stage by cryopreserving the sample, sectioning it transverse to the length of the construct and staining with hematoxylin and eosin or various fluorescent stains (Fig. 5b).
12. Just prior to implantation, wash the construct with PBS once.
13. While the construct is being anastomosed to the artery or vein, periodically irrigate the cavity and construct with sterile saline solution.

4. Notes

1. In place of performing bone marrow isolation, isolated MSCs are commercially available from various companies.
2. Commercially available media for MSCs is specifically designed with pre-screened FBS to keep cells undifferentiated. MSCGM should be used when expanding cells over multiple passages to prevent spontaneous differentiation. MSCGM consists of a basal media with supplements. Supplements are stored at −20°C and added to basal media, which is stored at 4°C. When the supplements are added to the basal media to complete the MSCGM formulation, the media is stored at 4°C and must be used within a month.
3. Although Percoll solution is most commonly used in MSC isolation, other density gradient solutions have been developed with densities of ~1.1 g/ml to separate mononuclear cells such as Ficoll-Paque and Histopaque-1077, available from Amersham Biosciences and Sigma, respectively. These solutions are ready-to-use and do not require high-speed centrifugation using a fixed-angle rotor.
4. Purchase antibodies that have been successfully tested for use in flow cytometry. CD29+ (integrin b1), CD44+ (receptor for hyaluronic acid and matrix proteins), CD105+ (endoglin, receptor for transforming growth factor-b(TGF-b) and integrins), CD166+ (a cell adhesion molecule), CD14- (monocyte surface antigen), CD34- (HSC surface antigen),

and CD45- (leukocyte surface antigen) are markers that Lonza (formerly Cambrex) uses to characterize MSCs; however, other markers can be used. MSCs have also been shown to express CD10, CD13, CD56, CD90 (24), CD73, and STRO-1 (a stromal cell surface antigen) (25). CD105 and CD73 have also been reported as SH2 and SH3, respectively (12, 26). In addition to cell surface markers, cells can be tested for their potential to differentiate into fat, bone, and cartilage to confirm that MSCs have been purified (12).

5. Alternatively, silane slides can be bought from Sigma. These slides have a frosted end that may be suboptimal for cell growth, depending on cell type.
6. HEPES buffer at pH 4 could be used for washing or collagen conjugation if it critical to avoid forming collagen polymer.
7. Any ECM with amine groups can be crosslinked to the gel with Sulfo-SANPAH, such as collagen and fibronectin.
8. Experiments can also be performed in round 100 mm dishes (need not be TC-treated), but since round dishes only fit one slide and also need 10 ml media to cover the slide, it is more economical to use 10 ml media for three slides in a square dish in order to save media as well as growth factors for growth factor experiments.
9. The weight must be optimized for the range of gel rigidities of interest. If it is too heavy, a soft gel may give an unstable reading and eventually flatten out to its maximum limit. If this happens use a lighter weight. Take the reading when the dial has stopped moving and gives a steady reading.
10. This step can be performed using adherence (Subheading 3.1.1) or density centrifugation (Subheading 3.1.2). Subheading 3.1.1 may give a more heterogeneous mixture of cells, but the majority of these cells express MSC markers. Subheading 3.1.2 may generate a more pure population. An alternative to both of these that is more specific is to purify MSCs using beads conjugated with specific antibodies for MSCs (available from Miltenyi Biotec). If you prefer to avoid this step altogether, isolated MSCs can be obtained from a commercial vendor (see Note 1).
11. Percoll is best used in fixed-angle rotors. However, it can also be used on swing-out rotors at 400×*g* for 20–30 min.
12. Although an end number of 10^4 cells are used for FACS analysis, about 10^5 cells are used initially to account for cell loss during centrifugation and washing steps.
13. Follow the recommended dilutions according to the company's product information sheet to empirically determine adequate amounts.

14. Slides are not directly placed on square dishes so that these dishes can be reused. Silane reacts with the plastic dishes and leave permanent residue. Parafilm is used since excess silane that remains on the surface can be disposed of.
15. "Back" of the slide refers to the slide that is not treated with silane. Washing the back of the slide gets rid of any silane that may have leaked onto that side.
16. One side of the glass is left unexposed to glutaraldehyde because the 24×60 mm coverslips that will be used to make the gel do not cover the entire glass slide. There will be 1.5 cm where the slide will not have gel, therefore glutaraldehyde need not be treated on this area. Also, spilling glutaraldehyde underneath slide is undesirable. These precautions are taken due to concerns of toxicity effects when the slide is exposed in the media during cell culture.
17. Acrylamide concentrations can be kept constant while changing bis acrylamide concentrations. For example, 6% acrylamide with a range of 0.5–0.02% bis-acrylamide concentrations can be used to vary rigidity.
18. Do not mark the slide directly but instead mark the dish. Using a marker can lead to cell death.
19. The purpose of the Gel Slick solution is to make the coverslip hydrophobic, thus allowing for easier removal of the coverslip from the gel after polymerization. Untreated coverslips are very difficult to remove from polymerized gel, especially the softer gels (less acrylamide/bis-acrylamide) that tend to be "sticky".
20. One end (the shorter edge) of coverslip is laid down first to help liquid travel faster over the entire slide.
21. This technique takes practice since it requires one to work fast before the gel polymerizes, while ensuring no bubbles being trapped and at the same time not allowing liquid to spill out from between the coverslip and the slide. Work gently, quickly, and carefully.
22. Be careful with softer gels since they stick to coverslip and small pieces of gel could potentially come off the slide. The key is lifting off the coverslip very slowly. For stiffer gels, although it is much easier to take coverslip off, be careful not to be too fast since the whole gel can come off with the coverslip.
23. Sulfo-SANPAH is a light and moisture-sensitive crosslinker that attaches amines of the acrylamide to the amines of collagen. Sulfo-SANPAH tends to stick to weigh paper so plastic weigh boat is optimal.
24. If edges are still wet, aspiration is an option but be *very careful* since it is easy to aspirate the gel, especially for the softer

gels since they hold more water and expand over the edges of the slide. Pouring out liquid from the dish is sufficient to avoid aspirating the gel, as long as the unexposed side of the slide is dry so the Sulfo-SANPAH solution will not leak into that area.

25. Photoactivated cross-linker is time-sensitive so it is important to act quickly until collagen is on the gel.
26. Longer incubation time is acceptable as long as gels do not dry out. If cell seeding will be done the next day, leaving slides in PBS overnight (continuing with steps 17–18) is a better option than leaving slides overnight in collagen, to avoid gels from drying out.
27. Do not leave UV light in TC hood on overnight to avoid degradation of gels. If UV light must be turned on, take dishes out, wrap in foil, and store in a clean place.
28. If cells are very confluent, or if lower confluency on gels is desired, use one round dish per 18–24 slides.
29. Aspirating 1.5 cm on one end of the slide will prevent the cell solution from leaking over the entire area of the control slide. Cells should be seeded on the same area as the slides with gels. Mark on the cell culture dish the side of the slide with the 1.5 cm space. Do not mark directly on the slide (see Note 18).
30. MSCGM can also be used if prevention of cell differentiation is desired. Make sure to break down pellet in a smaller volume before putting the full amount of media required.
31. Triplicates are recommended for statistical purposes. Gels are made in 6-wells instead of slides due to the requirements of the comparator setup.
32. A variety of polymers can be electrospun, including natural extracellular matrix molecules, such as collagen and elastin. The polymer concentration will have to be changed, along with the flow rate and needle/collection plate distance to produce quality nanofibers. The end-user should determine and characterize all conditions, as they are polymer-dependent.
33. It is not necessary to use MSCs when proceeding with the cell seeding step. Alternative cell types such as endothelial or smooth muscle cells can be used at varying seeding densities.

References

1. Bruder SP, Kurth AA, Shea M, Hayes WC, Jaiswal N, Kadiyala S (1998) Bone regeneration by implantation of purified, culture-expanded human mesenchymal stem cells. J Orthop Res 16:155–162
2. Ponticiello MS, Schinagl RM, Kadiyala S, Barry FP (2000) Gelatin-based resorbable sponge as a carrier matrix for human mesenchymal stem cells in cartilage regeneration therapy. J Biomed Mater Res 52:246–255

3. Toma C, Pittenger MF, Cahill KS, Byrne BJ, Kessler PD (2002) Human mesenchymal stem cells differentiate to a cardiomyocyte phenotype in the adult murine heart. Circulation 105:93–98
4. Ferrari G, Cusella-De Angelis G, Coletta M, Paolucci E, Stornaiuolo A, Cossu G, Mavilio F (1998) Muscle regeneration by bone marrow-derived myogenic progenitors. Science 279: 1528–1530
5. Mangi AA, Noiseux N, Kong D, He H, Rezvani M, Ingwall JS, Dzau VJ (2003) Mesenchymal stem cells modified with Akt prevent remodeling and restore performance of infarcted hearts. Nat Med 9:1195–1201
6. Brazelton TR, Rossi FM, Keshet GI, Blau HM (2000) From marrow to brain: expression of neuronal phenotypes in adult mice. Science 290:1775–1779
7. Crain BJ, Tran SD, Mezey E (2005) Transplanted human bone marrow cells generate new brain cells. J Neurol Sci 233:121–123
8. Caplan AI, Bruder SP (2001) Mesenchymal stem cells: building blocks for molecular medicine in the 21st century. Trends Mol Med 7:259–264
9. Wakitani S, Saito T, Caplan AI (1995) Myogenic cells derived from rat bone marrow mesenchymal stem cells exposed to 5-azacytidine. Muscle Nerve 18:1417–1426
10. Jiang Y, Jahagirdar BN, Reinhardt RL, Schwartz RE, Keene CD, Ortiz-Gonzalez XR, Reyes M, Lenvik T, Lund T, Blackstad M, Du J, Aldrich S, Lisberg A, Low WC, Largaespada DA, Verfaillie CM (2002) Pluripotency of mesenchymal stem cells derived from adult marrow. Nature 418:41–49
11. Galmiche MC, Koteliansky VE, Briere J, Herve P, Charbord P (1993) Stromal cells from human long-term marrow cultures are mesenchymal cells that differentiate following a vascular smooth muscle differentiation pathway. Blood 82:66–76
12. Pittenger MF, Mackay AM, Beck SC, Jaiswal RK, Douglas R, Mosca JD, Moorman MA, Simonetti DW, Craig S, Marshak DR (1999) Multilineage potential of adult human mesenchymal stem cells. Science 284:143–147
13. Prockop DJ (1997) Marrow stromal cells as stem cells for nonhematopoietic tissues. Science 276:71–74
14. Bartholomew A, Sturgeon C, Siatskas M, Ferrer K, McIntosh K, Patil S, Hardy W, Devine S, Ucker D, Deans R, Moseley A, Hoffman R (2002) Mesenchymal stem cells suppress lymphocyte proliferation in vitro and prolong skin graft survival in vivo. Exp Hematol 30:42–48
15. Krampera M, Glennie S, Dyson J, Scott D, Laylor R, Simpson E, Dazzi F (2003) Bone marrow mesenchymal stem cells inhibit the response of naive and memory antigen-specific T cells to their cognate peptide. Blood 101: 3722–3729
16. Devine SM, Cobbs C, Jennings M, Bartholomew A, Hoffman R (2003) Mesenchymal stem cells distribute to a wide range of tissues following systemic infusion into non-human primates. Blood 101:2999–3001
17. Erices A, Conget P, Minguell JJ (2000) Mesenchymal progenitor cells in human umbilical cord blood. Br J Haematol 109:235–242
18. Zvaifler NJ, Marinova-Mutafchieva L, Adams G, Edwards CJ, Moss J, Burger JA, Maini RN (2000) Mesenchymal precursor cells in the blood of normal individuals. Arthritis Res 2:477–488
19. Miao Z, Jin J, Chen L, Zhu J, Huang W, Zhao J, Qian H, Zhang X (2006) Isolation of mesenchymal stem cells from human placenta: comparison with human bone marrow mesenchymal stem cells. Cell Biol Int 30:681–687
20. Gronthos S, Franklin DM, Leddy HA, Robey PG, Storms RW, Gimble JM (2001) Surface protein characterization of human adipose tissue-derived stromal cells. J Cell Physiol 189:54–63
21. Friedenstein AJ, Piatetzky S II, Petrakova KV (1966) Osteogenesis in transplants of bone marrow cells. J Embryol Exp Morphol 16: 381–390
22. Haynesworth SE, Goshima J, Goldberg VM, Caplan AI (1992) Characterization of cells with osteogenic potential from human marrow. Bone 13:81–88
23. Li D, Xia Y (2004) Electrospinning of nanofibers: reinventing the wheel? Adv Mater 16:1151–1170
24. Young HE, Steele TA, Bray RA, Detmer K, Blake LW, Lucas PW, Black AC Jr (1999) Human pluripotent and progenitor cells display cell surface cluster differentiation markers CD10, CD13, CD56, and MHC class-I. Proc Soc Exp Biol Med 221:63–71
25. Tuli R, Tuli S, Nandi S, Wang ML, Alexander PG, Haleem-Smith H, Hozack WJ, Manner PA, Danielson KG, Tuan RS (2003) Characterization of multipotential mesenchymal progenitor cells derived from human trabecular bone. Stem Cells 21:681–693
26. Bartholomew A, Patil S, Mackay A, Nelson M, Buyaner D, Hardy W, Mosca J, Sturgeon C, Siatskas M, Mahmud N, Ferrer K, Deans R, Moseley A, Hoffman R, Devine SM (2001) Baboon mesenchymal stem cells can be genetically modified to secrete human erythropoietin in vivo. Hum Gene Ther 12:1527–1541

Chapter 9

Preparation and Analysis of Endothelial Progenitor Cells Associated with Angiogenesis

Steve Khachi, Yiqian Zhu, and Guo-Yuan Yang

Abstract

Endothelial progenitor cell-based therapy offers great potential to reduce morbidity and mortality in patients with vascular diseases. In animal models of ischemia, circulating endothelial progenitor cells (EPCs) have been shown to home to sites of active angiogenesis and differentiate into endothelial cells in response to tissue ischemia, vascular trauma, or tumor growth. These studies indicate a thorough understanding of EPCs function and role in angiogenesis as a potential therapeutic target for vascular diseases. In this chapter, we describe the procedures for: (1) isolation of EPCs from circulating blood and bone marrow; (2) culturing these cells; (3) characterization of EPCs by FCAS, DiI-ac-LDL up-taking, and FITC-UEA-1 binding assays; and (4) determination of EPCs' function by examining proliferation, migration, and tube formation in vitro.

Key words: Endothelial progenitor cells, Angiogenesis, Bone marrow

1. Introduction

Neovascularization, the generation of new blood vessels, is made possible through vasculogenesis, arteriogenesis, or angiogenesis. Vasculogenesis is a de novo process, in which progenitor stem cells give rise to new vascular networks. This form of neovascularization plays a significant role in the embryological development of the vascular system. Arteriogenesis is a process in which the diameter of an existing arterial vessel increases in order to redirect blood flow to areas of higher demand. Angiogenesis is a step-by-step process that leads to sprouting of new microvessels from pre-existing microvessels, and appears to be the main pathway of new blood vessel formation in adults. Angiogenesis involves concerted actions of many angiogenic molecules leading to increases of vascular permeability, degradation of surrounding matrix, pro-

Irina M. Conboy et al. (eds.), *Protocols for Adult Stem Cells*, Methods in Molecular Biology, vol. 621,
DOI 10.1007/978-1-60761-063-2_9, © Springer Science+Business Media, LLC 2010

liferation and migration of endothelial cells, and stabilization of neomicrovessels (1). Adult circulating endothelial progenitor cell (EPC) appear to play a significant role in angiogenesis, resulting in repair and recovery after ischemic injuries (2–6).

EPC, first reported by Asahara in 1997, were initially identified from human peripheral blood through the expression of CD34 or CD133, the more immature human stem cell (HSC) marker protein (7). Both CD34-positive and CD133-positive cell populations differentiated to ECs in vitro under appropriate endothelial differentiation-promoting conditions (8), according to the initial discovery.

EPCs were defined as cells that were positive for both Human stem cell (HSC) markers (e.g., CD34 or CD133), and endothelial cell markers (e.g., KDR, vWF, VE-Cadherin, Tie2, and CD31 positive staining) (9, 10). The sources of EPCs include adult bone marrow (11), peripheral blood (12, 13), cord blood (14), and tissue resident cells (HSC, MSC, cKit+, neuronal progenitor cells). Circulating EPCs appear to be particularly important because they directly contribute to endothelial recovery, or rapidly migrate into the region of activating angiogenesis.

Ex vivo expansion of EPCs can be derived from peripheral blood mononuclear cell populations. When endothelial mitogens are incubated with vascular endothelial growth factor, basic fibroblast growth factor, insulin-like growth factor, and epidermal growth factor for 7–10 days, these EPCs expand 80- to 90-fold (7, 15, 16). Therefore, at least two types of EPCs exist: one type has a spindle-like shape, which is termed early EPC; another type has a cobblestone appearance, which is called late EPC (11, 12). Although late EPCs share phenotypes with mature ECs, they may proliferate much faster and preserve stem cell characteristics. Cultured late EPCs are now being used in experiments for the following reasons: (1) the number of EPCs obtained by ex vivo expansion exceeds the number of CD34-positive cells that can be freshly isolated; (2) the purity and quality of expended EPCs in a cultured population are superior to those of freshly isolated CD34-positive cells; and (3) for therapeutic strategies designed to employ transplanted cells, which constitutively express pro- or anti-angiogenic factors, gene transfer of EPCs is facilitated by the use of culture-committed late EPCs vs. less differentiated early EPCs. Early EPCs contribute to vasculogenesis, mainly by secreting angiogenic cytokines (e.g., VEGF, FGF, EGF, IGF, granulocyte-colony-stimulating factor), while late EPCs enhance vasculogenesis by providing sufficient numbers of endothelial cells based on their high proliferation potency (17).

Many physiological/pathological conditions or drugs influence the number of circulating EPCs (18–20). The number is reduced in patients with risk factors for ischemic cardiovascular diseases (20). Acute myocardial infarction, coronary bypass grafting, or burn injury rapidly increases circulating EPCs. Using laser-scanning confocal microscopy, implanted peripheral blood CD34-positive cells have been found to differentiate into glial cells (GFAP positive), neurons (Nestin, MAP-2 or NeuN positive), and vascular endothelial cells (vWF positive), thereby enhancing neuroplastic effects in the ischemic brain (21). Mature mouse ECs from embryonic stem cells contribute to activating angiogenesis (22). Furthermore, CD133 positive angioblasts derived from human embryonic stem cells can differentiate into mature human ECs and contribute to activating angiogenesis in immuno-compromised mice (23). These findings indicate that circulating endothelial progenitor cells can provide a novel therapeutic strategy for tissue ischemia and vascular diseases by contributing to postnatal angiogenesis and other regenerative pathways.

2. Materials

2.1. EPC Isolation and Culture

2.1.1. EPC from Circulating Blood

1. Heparin (Immuno, Vienna, Austria).
2. Phosphate Buffered Saline (PBS), without magnesium and calcium, pH 7.4 (Sigma, St. Louis, MO).
3. Histpaque 1077 (Sigma-Aldrich, St Louis, MO).
4. Fetal bovine serum (FBS, Gibco, Carlsbad, CA). Store at −20°C.
5. EBM-2 MV Bullet kit system (Clonetics, San Diego, CA).
6. Pasteur pipettes.
7. SigmaScreen™ Ffibronectin coated multiwell plate (Sigma-Aldrich, St Louis, MO).
8. 6-well plate.
9. Trypsin/0.25% EDTA (Gibco, Carlsbad, CA). Store at −20°C. When ready to use, dilute to 0.5× in PBS and store at 4°C.

2.1.2. EPC from Bone Marrow Cell

1. 23G needle.
2. Heparin (Immuno, Vienna, Austria).
3. Deoxyribonuclease I (Sigma, St. Louis, MO).
4. Dulbecco's phosphate buffer saline (PBS, PAA Laboratories, Austria).
5. Lymphoprep density gradient (Nycomed, Oslo, Norway).
6. Fetal calf serum (FCS, PAA Laboratories, Austria).
7. Rat-derived fibronectin Powder (Sigma, St. Louis, MO).
8. 12-well plate.

9. EBM-2 MV Bullet kit system (Clonetics, San Diego, CA).
10. Stem cell growth factor (Prepro-Tech EC Ltd, USA).
11. Type CLS-CI-22 Collagenase (Biochrom AG, Berlin, Germany).

2.2. EPC Characterization

2.2.1. FCAS Assay

1. PBS, without magnesium and calcium, pH 7.4.
2. FcR blocking reagent (Miltenyi Biotec, Cat. No, 130-059-901).
3. Red blood cells lysing buffer (Sigma, St. Louis, MO).
4. 0.2% bovine serum albumin (BSA) in PBS (Sigma, St. Louis, MO).
5. 2% paraformaldehyde in PBS (Fisher Scientific, Pittsburgh, PA).
6. Antibodies against CD34, CD133, VEGFR2 (see Note 3).
7. Isotype identical antibodies as negative control (BD Bioscience, San Diego, CA, USA).
8. Flow cytometer.

2.2.2. Dil-ac-LDL Up-taking Assay

1. EBM-2 MV Bullet kit system (Clonetics, San Diego, CA).
2. Trypsin/0.25% EDTA (Gibco, Carlsbad, CA). Store at −20°C. When ready to use, dilute to 0.5× in PBS and store at 4°C.
3. DiI cell-labeling solution (Invitrogen, Carlsbad, CA).

2.2.3. FITC-UEA-1 Binding Assay

1. DiI-ac-LDL (Molecular Probes, Eugene, OR, USA).
2. PBS, without magnesium and calcium, pH 7.4.
3. 2% paraformaldehyde in PBS.
4. Goat serum (Sigma, St. Louis, MO).
5. Fluorescein isothiocyanate-labeled Ulex europaeus agglutinin-1 (Sigma, St. Louis, MO).
6. 4′,6-Diamidino-2-phenylindole (Invitrogen, Carlsbad, CA).

2.3. Determination of the Function of EPC In Vitro

2.3.1. EPC Proliferation

1. EBM-2 MV Bullet kit system (Clonetics, San Diego, CA).
2. 0.1% BSA in PBS.
3. Bromodeoxyuridine kit (Roche, Mannheim, Germany).
4. Anti-BrdU antibody (Abcam, Cambridge, UK).
5. ELISA reader.

2.3.2. EPC Migration

1. 0.1% BSA in PBS.
2. 8.0 mm pore polycarbonate filter inserts (Costar, San Diego, CA, USA).
3. Hematoxylin (Sigma, St. Louis, MO).

2.3.3. EPC Tube Formation

1. Growth factor-reduced matrigel (BD Bioscience, San Diego, CA, USA).
2. 0.1% BSA in PBS.
3. EBM-2 MV Bullet kit system (Clonetics, San Diego, CA).
4. PBS, without magnesium and calcium, pH 7.4.

3. Methods

3.1. EPC Isolation and Culture

3.1.1. EPC from Circulating Blood

1. Collect at least 20 ml of peripheral blood using an anti-coagulant (see Note 1) to avoid clotting or clumping of the sample.
2. Add peripheral blood to a 50 ml polypropylene tube containing 10 ml of PBS and mix gently.
3. Add 10 ml Histpaque 1077 into 2 × 50 ml tube individually.
4. Gently overlay 15 ml of the diluted blood on top of the Histpaque. Do not break the interface.
5. Centrifuge tubes at 400 × *g* for 30 min at room temperature with the brake turned to the "off" position (see Note 2).
6. After centrifugation, carefully collect the mononuclear cells (MNCs), which are between the Histpaque and plasma layers, and transfer to a new 50 ml polypropylene tube. Add PBS with 2% FBS to bring up the final volume to 40 ml.
7. Centrifuge at 250 × *g* with "brake and acceleration" for approximately 10 min at room temperature to pellet the cells.
8. Decant supernatant and resuspend the pellet with 6 ml of PBS with 2% FBS.
9. Transfer cell suspension to a new 15 ml centrifuge tube. Rinse the 50 ml tube with an additional 6 ml of PBS with 2% FBS.
10. Centrifuge 10 min at room temperature with "brake and acceleration" to pellet the cells.
11. Decant supernatant and re-suspend pellet in 1–3 ml (depending on size of the cell pellet) of EBM2 + MV single quote medium. Count nucleated cells with 1/20 dilution.
12. Seed the MNCs onto fibronectin coated 3.5 cm plates at 1×10^7 cell per well (6-well plate) or at 0.8×10^7 per 3.5 cm plate.
13. Incubate for 3 days at 37°C, 5% CO_2 with ≥95% humidity.
14. Carefully and gently change the medium with fresh EBM-2 MV. Do not remove the suspended cells. Keep on culturing for another 2 days.

15. After 3–5 days, the attached cell colonies, which are proven to be both progenitor and endothelial cell marker-positive, are named early EPCs.
16. Change medium and remove the suspended cells. Keep on culturing and change medium every 3 days in EBM2 + MV.
17. After another 5–6 days, cell colonies, namely the expended EPCs, will be produced.
18. Trypsinize and reseed the cells onto 0.25% Gelatin coated plate in EBM2 + MV single-quote medium. The cells are defined to be P1 and can be passaged for several generations.

3.1.2. EPC from Bone Marrow Cell

1. Aspirate bone marrow cells by a 23G needle from rat's femur.
2. Mix the aspirated bone marrow with 1,000 U/ml heparin, and 1,000 U/ml deoxyribonuclease I in PBS.
3. Obtain mononuclear cell fraction from a Lymphoprep density gradient after centrifugation for 30 min at 250 × *g*.
4. Card the mononuclear cell fraction, then wash and centrifuge it at 75 × *g* for 10 min.
5. Suspend the cell pellet in EBM-2 medium, and supplement with 20% FCS.
6. Plate the suspended cell pellet onto rat-derived fibronectin-coated (10 μg/ml) 12-well plate for 24 h.
7. Aspirate and transfer the non-adherent cell population to a new fibronectin-coated plate for another 24 h.
8. Repeat procedure 7 to remove the rapidly adherent hematopoietic cells or mature ECs.
9. Harvest only the non-adherent cell population and culture it in EBM-2 medium containing VEGF, *h*FGF-B, R3-IGF-1, *h*EGF, ascorbic acid, hydrocortisone, gentamycin, amphotericin B, and stem cell growth factor (SCGF).
10. After 2–3 days, angioblasts-like cells will appear, and spindle-shaped cell outgrowth can be observed.
11. After 7–10 days, confluence of the outgrowing cell population will be reached, and cells can be divided by collagenase (Type CLS-CI-22).

3.2. EPC Characterization

3.2.1. Fluorescence-activated Cell Sorting

1. Wash 4 ml of peripheral blood by adding 200 ml of PBS, and pelleting cells by centrifuging the plate carriers at 150 × *g* for 2 min. Discard the supernatant.
2. Incubate cells in PBS and block by FcR blocking reagent for 15 min to block Fc receptors.
3. Incubate cells with mouse anti-human monoclonal antibodies for 30 min at 4°C in the dark.

4. Lyse red blood cells (RBC) by RBC lysing buffer, and wash by two cycles of centrifuging at 200 × *g* and re-suspending in PBS with 0.2% BSA.
5. Subsequently, re-suspend cells in 2% paraformaldehyde.
6. Analyze the cells by fluorescence-activated cell sorting (FACS) and collect 100,000 events per sample.

3.2.2. Dil-ac-LDL Up-taking Assay

1. Trypsinze and re-suspend cultured *h*EPCs in 3–5 passages with 80% confluence at a density of 1×10^6/ml in EBM 2 medium.
2. Add 5 μl of DiI cell-labeling solution per ml of cell suspension, and incubate for 20 min at 37°C.
3. Spin the labeled cells down at 200 × *g* for 5 min.
4. Wash the cell pellet twice with EBM2.
5. Re-suspend cells in 200 μl of EBM2 if cells are prepared for injection in vivo.

3.2.3. FITC-UEA-1 Binding Assay

1. Incubate *h*EPCs in chamber slides with a medium containing 2.5 μg/ml of DiI-ac-LDL for 1 h at 37°C.
2. Wash cells with PBS 3 time.
3. Then fix cells with 2% paraformaldehyde for 10 min.
4. Wash with PBS 3 times.
5. Block cells by using 0.2% goat serum in PBS for 1 h.
6. Subsequently, incubate cells with fluorescein isothiocyanate-labeled Ulex europaeus agglutinin-1 (UEA-1, 10 μg/ml) at room temperature for 1 h.
7. Wash cells with PBS 3 times to remove free UEA-1.
8. Counterstain cells with 4′, 6-diamidino-2-phenylindole (DAPI).
9. Cells will be ready to be examined under a fluorescence microscope.

3.3. Determination of the Function of EPC In Vitro

3.3.1. EPC Proliferation

1. Seed endothelial progenitor cells onto a 96-well gelatin-coated plate (1×10^4) with EBM-2/Single Quote medium and culture up to 75% confluence.
2. Incubate cells overnight in EBM-2 with 0.1% BSA.
3. Label EPCs with bromodeoxyuridine (BrdU) overnight.
4. Apply alkaline phosphatase-conjugated anti-BrdU antibody for 90 min.
5. Measure absorbance at 450 nm by using an ELISA reader after 10 min of incubation in subject buffer.
6. Analyze results as the mean percentage of OD450 over control groups.

3.3.2. EPC Migration

1. Suspend serum-starved EPCs in medium with 0.1% BSA at a concentration of 2.5×10^5 ml using a 24-well transwell cell culture chamber with 8.0 mm pore polycarbonate filter inserts.
2. Apply cells to precoated insert filters.
3. Add test factors in serum-free medium and place in the lower chamber.
4. Incubate cells at 37°C for 18 h.
5. Scrape the cells on the upper side of the filter, so the migrated cells on the lower side will be fixed.
6. Stain the fixed cells with hematoxylin, and mount the membrane on a slide in order to examine under microscope.
7. Migrated cells can be quantified by measuring the stained cells in five random areas per membrane, and the results can be calculated by the mean percentage of migrating cells over control groups.

3.3.3. EPC Tube Formation

1. Coat a 96-well culture plate with 50 μl of growth factor-reduced matrigel per well, and allow polymerization for 30 min at 37°C.
2. Seed pretreated cells on the matrigel in 0.1% BSA/EBM-2 medium with testing factors and incubate for 18 h at 37°C.
3. Wash with PBS, and then capture images with a light microscope.
4. Quantify tube formation ability by counting the total number of cell cluster and branching under 3 × 4 magnified fields per well. The results should be expressed as mean fields of cell clusters and branching compared to the control groups.

4. Notes

1. Sodium heparin, ACD (acid citrate dextrose), K3 EDTA, or equivalent may be used as an anti-coagulant agent.
2. There should be neither brake nor acceleration during centrifugation.
3. The antibodies to be used are R-Phycoerythrin (PE) conjugated anti-human-CD34 (BD Biosciences), PE conjugated anti-human CD133 (Miltenyi Biotec), and Allophycocyanin (APC) conjugated anti-human VEGFR2 (KDR, R&D systems, Minneapolis, MN). Isotype identical antibodies will serve as controls. The data will be analyzed by Cellquest (BD Biosciences) and expressed as the percentage of EPC cells in Peripheral Blood Mononuclear Cells (PBMC).

References

1. Conway EM, Collen D, Carmeliet P (2001) Molecular mechanisms of blood vessel growth. Cardiovasc Res 49:507–521
2. Shintani S, Murohara T, Ikeda H, Ueno T, Honma T, Katoh A, Sasaki K, Shimada T, Oike Y, Imaizumi T (2001) Mobilization of endothelial progenitor cells in patients with acute myocardial infarction. Circulation 103:2776–2779
3. Ii M, Nishimura H, Iwakura A, Wecker A, Eaton E, Asahara T, Losordo DW (2005) Endothelial progenitor cells are rapidly recruited to myocardium and mediate protective effect of ischemic preconditioning via "imported" nitric oxide synthase activity. Circulation 111:1114–1120
4. Kalka C, Masuda H, Takahashi T, Kalka-Moll WM, Silver M, Kearney M, Li T, Isner JM, Asahara T (2000) Transplantation of ex vivo expanded endothelial progenitor cells for therapeutic neovascularization. Proc Natl Acad Sci USA 97:3422–3427
5. Kawamoto A, Tkebuchava T, Yamaguchi J, Nishimura H, Yoon YS, Milliken C, Uchida S, Masuo O, Iwaguro H, Ma H, Hanley A, Silver M, Kearney M, Losordo DW, Isner JM, Asahara T (2003) Intramyocardial transplantation of autologous endothelial progenitor cells for therapeutic neovascularization of myocardial ischemia. Circulation 107:461–468
6. Szmitko PE, Fedak PW, Weisel RD, Stewart DJ, Kutryk MJ, Verma S (2003) Endothelial progenitor cells: new hope for a broken heart. Circulation 107:3093–3100
7. Asahara T, Murohara T, Sullivan A, Silver M, van der Zee T, Li T, Witzenbichler B, Schatteman G, Isner JM (1997) Isolation of putative progenitor endothelial cells for angiogenesis. Science 275:964–967
8. Gehling UM, Ergun S, Schumacher U, Wagener C, Pantel K, Otte M, Schuch G, Schafhausen P, Mende T, Kilic N, Kluge K, Schafer B, Hossfeld DK, Fiedler W (2000) In vitro differentiation of endothelial cells from AC133-positive progenitor cells. Blood 95:3106–3112
9. Miraglia S, Godfrey W, Yin AH, Atkins K, Warnke R, Holden JT, Bray RA, Waller EK, Buck DW (1997) A novel five-transmembrane hematopoietic stem cell antigen: isolation, characterization, and molecular cloning. Blood 90:5013–5021
10. Quirici N, Soligo D, Caneva L, Servida F, Bossolasco P, Deliliers GL (2001) Differentiation and expansion of endothelial cells from human bone marrow CD133(+) cells. Br J Haematol 115:186–194
11. Reyes M, Dudek A, Jahagirdar B, Koodie L, Marker PH, Verfaillie CM (2002) Origin of endothelial progenitors in human postnatal bone marrow. J Clin Invest 109:337–346
12. Lin Y, Weisdorf DJ, Solovey A, Hebbel RP (2000) Origins of circulating endothelial cells and endothelial outgrowth from blood. J Clin Invest 105:71–77
13. Peichev M, Naiyer AJ, Pereira D, Zhu Z, Lane WJ, Williams M, Oz MC, Hicklin DJ, Witte L, Moore MA, Rafii S (2000) Expression of VEGFR-2 and AC133 by circulating human CD34(+) cells identifies a population of functional endothelial precursors. Blood 95: 952–958
14. Murohara T, Ikeda H, Duan J, Shintani S, Sasaki K, Eguchi H, Onitsuka I, Matsui K, Imaizumi T (2000) Transplanted cord blood-derived endothelial precursor cells augment postnatal neovascularization. J Clin Invest 105:1527–1536
15. Nishikawa SI, Nishikawa S, Hirashima M, Matsuyoshi N, Kodama H (1998) Progressive lineage analysis by cell sorting and culture identifies FLK1 + VE-cadherin + cells at a diverging point of endothelial and hemopoietic lineages. Development 125:1747–1757
16. Shi Q, Rafii S, Wu MH, Wijelath ES, Yu C, Ishida A, Fujita Y, Kothari S, Mohle R, Sauvage LR, Moore MA, Storb RF, Hammond WP (1998) Evidence for circulating bone marrow-derived endothelial cells. Blood 92:362–367
17. Hur J, Yoon CH, Kim HS, Choi JH, Kang HJ, Hwang KK, Oh BH, Lee MM, Park YB (2004) Characterization of two types of endothelial progenitor cells and their different contributions to neovasculogenesis. Arterioscler Thromb Vasc Biol 24:288–293
18. Llevadot J, Murasawa S, Kureishi Y, Uchida S, Masuda H, Kawamoto A, Walsh K, Isner JM, Asahara T (2001) HMG-CoA reductase inhibitor mobilizes bone marrow-derived endothelial progenitor cells. J Clin Invest 108:399–405
19. Walter DH, Rittig K, Bahlmann FH, Kirchmair R, Silver M, Murayama T, Nishimura H, Losordo DW, Asahara T, Isner JM (2002) Statin therapy accelerates reendothelialization: a novel effect involving mobilization and incorporation of bone marrow-derived endothelial progenitor cells. Circulation 105:3017–3024
20. Werner N, Priller J, Laufs U, Endres M, Bohm M, Dirnagl U, Nickenig G (2002) Bone marrow-derived progenitor cells modulate

vascular reendothelialization and neointimal formation: effect of 3-hydroxy-3-methylglutaryl coenzyme a reductase inhibition. Arterioscler Thromb Vasc Biol 22:1567–1572

21. Shyu WC, Lin SZ, Chiang MF, Su CY, Li H (2006) Intracerebral peripheral blood stem cell (CD34+) implantation induces neuroplasticity by enhancing beta1 integrin-mediated angiogenesis in chronic stroke rats. J Neurosci 26:3444–3453

22. Marchetti S, Gimond C, Iljin K, Bourcier C, Alitalo K, Pouyssegur J, Pages G (2002) Endothelial cells genetically selected from differentiating mouse embryonic stem cells incorporate at sites of neovascularization in vivo. J Cell Sci 115:2075–2085

23. Levenberg S, Golub JS, Amit M, Itskovitz-Eldor J, Langer R (2002) Endothelial cells derived from human embryonic stem cells. Proc Natl Acad Sci USA 99:4391–4396

Chapter 10

Preparation of Adult Muscle Fiber-Associated Stem/Precursor Cells

Michael J. Conboy and Irina M. Conboy

Abstract

In our studies of muscle regeneration we have developed, modified, and optimized techniques to isolate and study the stem and precursor cells to muscle tissue. Our goals have been to obtain for study muscle fibers in bulk, or the fiber-associated cells, separately from the other cells found in muscle. Using these techniques, myofiber-associated cells may be isolated from neonatal through adult muscle, from resting or from regenerating muscle, thus allowing one to investigate the cellular populations participating during the time course of these events. The protocol is applicable to any age and condition of muscle and may be adapted for other tissues.

Key words: Satellite cell, Myoblast, Myofiber, Hepatocyte, Muscle, Myogenic, Myogenesis, Adult stem cell

1. Introduction

The primary organ stem and progenitor cells from a tissue differ in many respects from immortal cell lines that may be cultured out of such tissue. The growth potential of immortal lines makes them invaluable for many genetic and biochemical experiments, but at some point in the study of the development or regeneration of the tissue, one must return to study the native, primary cell. From any given tissue, obtaining stem/precursor cells in sufficient numbers and at sufficient purity is a prerequisite for further study. In muscle, the stem cell pool resides in the satellite cell position on the muscle fiber (myofiber) itself, a position defined as between the basal lamina extracellular matrix and the plasma membrane of the multinucleated fiber cell (Fig. 1). Satellite cells are typically mitotically quiescent until activated for muscle growth or regeneration, whereupon they can account for virtually all the muscle

Irina M. Conboy et al. (eds.), *Protocols for Adult Stem Cells*, Methods in Molecular Biology, vol. 621,
DOI 10.1007/978-1-60761-063-2_10, © Springer Science+Business Media, LLC 2010

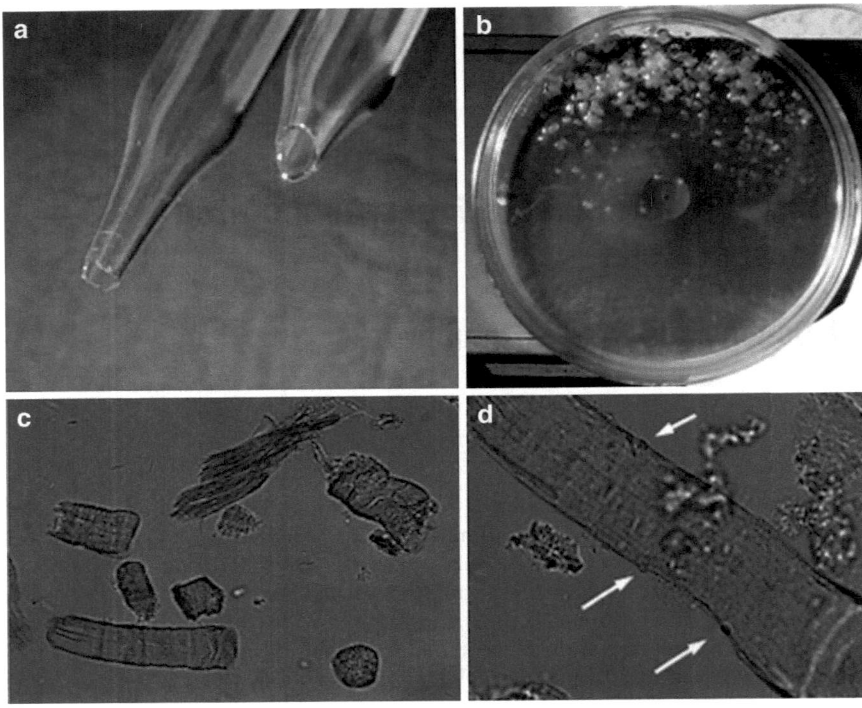

Fig. 1. (**a**) Image of Pasteur pipettes with the tip broken off. Openings of this size are used to triturate gram amounts of small muscle. (**b**) Image of tilting a 10 cm dish to drain the dissociated fibers from the remaining muscle and connective tissue, described in Subheading 3.4 above. (**c**) Brightfield image of muscle fibers prepared in PBS without serum (see Subheading 3.7 and Note 4). Note the shredded fiber fragments and the many tiny myofibrils. (**d**) Brightfield image of fibers prepared with 10% serum in the medium. Arrows point to satellite cells associated with the fibers. The diameter of the fiber is approximately 100 um

fiber regenerative capacity (1). Additionally, in the muscle tissue, there are a host of other cell types: between the fibers are poorly defined interstitial cells which appear to include myogenic, adipogenic, and fibroblastic cells, and among the myofiber bundles are the endothelial blood vessel cells, the connective tissue maintenance cells, especially rich at tendons, the nerve endings and their support cells, resident and itinerant immune, inflammatory and dendritic cells, lipid-laden adipocytes, and the surrounding fascia tissue.

A number of markers may be used to identify these different cell types and to distinguish different stages of differentiation during muscle regeneration. Quiescent satellite cells express the cell surface markers CD34, m-Cadherin and Pax7 (2, 3), but low to undetectable levels of many other distinguishing markers; the Myf-5 locus appears to be transcribed but many researchers have failed to detect MYF5 protein (including the authors). Upon activation, satellite cells (and the injured myofiber), express Delta and activate the Notch pathway, as demonstrated by the presence of cleaved nuclear Notch-1. Notch signals induce these cells to begin to proliferate, and upregulate Pax3 (4). At this point, particularly if the fibers have been injured, the satellite cell position may be histologically unclear, and we refer to the proliferating myogenic cell population as intermediate proliferative, myogenic precursor cells. At least two to three populations of myofiber-associated cells can be distinguished here: one is Sca1 negative, will upregulate Pax7, Myf5 protein and MyoD, is desmin-high and capable of fusion into myotubes, defining that population as myoblasts (Figs. 2 and 3). Upon differentiation and/or fusion with myotubes, the myoblasts will express the terminal differentiation program including myogenin and eMHC and still continue to express desmin (Figs. 2 and 3). One to two weeks after new myotubes are produced by myoblasts, eMHC and desmin are down-regulated and these markers are not detectable in mature myofibers (4–6). The second one is Sca1+ve, desmin-low population, and is not autonomously myogenic, but may generate muscle marker expressing cells upon coculture with Sca1 negative cells. In contrast to Sca1 negative cells, these Sca1+ve cells give rise to both fibroblasts and adipocytes with high efficiency (7, 8). These populations appear to be present throughout muscle homeostasis, injury, and regeneration (9). Renewing satellite cells may come from both populations, and certainly from the first (Pax 7+ve) pool (8, 10). Pax 7 is also expressed at high levels by fusion-competent, desmin+ve myoblasts (Fig. 2). From 2 days through 5 days after injury in the adult mouse, leukocytes expressing the CD45 antigen infiltrate the injury site and the damaged myofibers themselves, increasing the population of mononucleated cells to a size at least as large as the myogenic precursors (11, 12). These infiltrating CD45+ cells do not contribute appreciably to new muscle tissue, but may elaborate factors that modulate the myogenic function of resident muscle precursors.

There are two general approaches for obtaining muscle precursor cells from fresh muscle tissue for study. The first is to obtain all the cells from the muscle by mincing and enzymatically digesting the whole tissue, which permits the analysis of a large number of mononuclear cells from developing, quiescent or regenerating muscle, and provides a source for primary cell culture. However, this method yields a very heterogenous cell population: the muscle precursor cells associated with the fiber along with interstitial cells, both myogenic and endothelial, the ubiquitous fibroblasts and a varying proportion of blood cells, depending on whether the muscle was uninjured (low frequency of blood lineage cells) or regenerating (high frequency of blood lineage cells). Bischoff (13), developed a second protocol describing the isolation of whole, single muscle fibers with attached satellite cells (satellite by definition), without significant contamination of other cell types. This and derived protocols

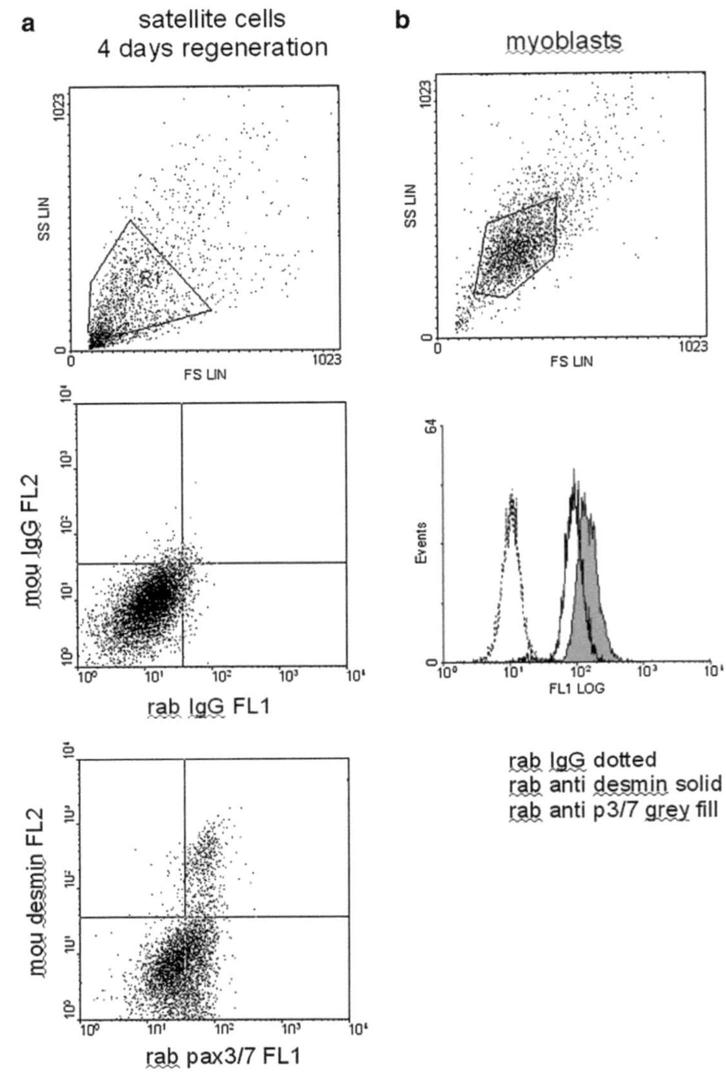

Fig. 2. (**a**) Mouse muscle was processed by the above methods, using the Variation 3.9.3 on the live cells analyzed by flow cytometry. Satellite and myofiber-associated cells are characteristically small and nongranular by forward and side scatters, and so this population is typically recognizable in dot or contour plots, even with a large amount of debris. Variation 3.9.2 typically gives too much debris for FSC and SSC gating, without some further purification step. Satellite cells from regenerating muscle (**a**) and long-term cultured primary myoblasts produced by such satellite cells in vitro (**b**) were immunostained as in Chapter 11, for the myogenic cell markers: rabbit anti-Pax3/Pax7 (using an antibody cross-reactive for these closely homologous proteins) and desmin, and analyzed by FACS. Desmin+ve cells are always Pax3/Pax7+ve and all primary myoblasts coexpress these myogenic markers

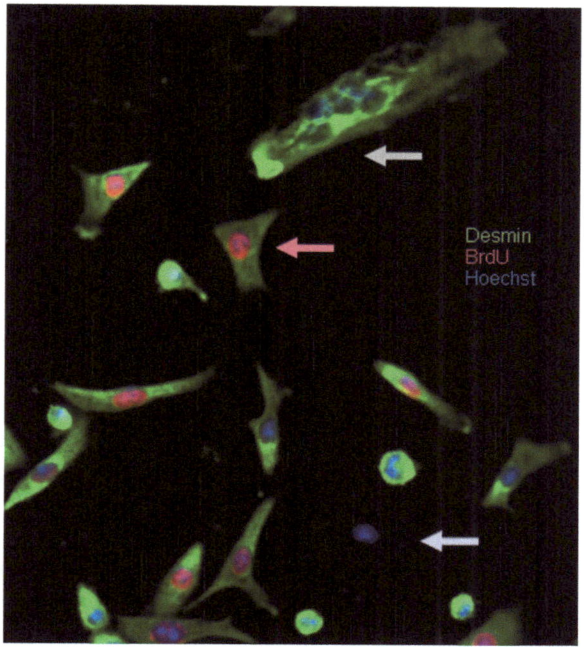

Fig. 3. Myofibers with associated satellite cells were prepared by the above protocols, Variation 3.7.3, from regenerating muscle and cultured in suspension on ECM-coated tissue culture plates in Ham's F10 with 20% fetal bovine serum and 5 ng/ml bFGF for 24 h with BrdU administered for the last 2 h. Based on immunostaining, multitudes of myogenic cells migrated from these myofibers onto the ECM-coated culture plates and formed colonies of proliferating fusion-competent desmin+ve myoblasts (*pink arrow*) and de-novo desmin+ve myotubes (*green arrow*). Hoechst (*blue*) labels all nuclei. Few non-myogenic cells (*blue arrow*) are typically present in these myofiber cultures. Similar results are obtained from culturing purified satellite cells using the Variations 3.9, although less total myogenic cells are produced as compared with culturing the bulk myofiber cultures prepared with variation 3.7.3

helped establish the "single fiber" model, which has supported the hypothesis that the fiber with associated satellite cells is the unit of regeneration in muscle (1). We sought a method to obtain the cell populations from quiescent muscle or at any point during regeneration, in a substantial quantity that would be on par with the first protocol, yet of a purity approaching that of the second.

We routinely achieve 50 to 1,000-fold enrichment from the original muscle depending on the variation of the methods below, (e.g., 1 g of muscle gives 1–20 μl of cells + debris). Some satellite or myofiber associated cells are invariably lost to the "interstitial" fraction, but few interstitial cells are carried along with the fiber fraction. For cell numbers, expect 4×10^4 cells per gram of resting muscle (from adult C57 mice), and up to an order of magnitude more from regenerating muscle. Gated by forward and side-scatter characteristic of small, nongranular cells, the gated cells are >95% CD34+ve/m-Cadherin+ve and less than 2% PECAM+ endothelial cells (6).

From uninjured muscle or with negative selection against CD45+ve cells, less than 1% of the mononucleate cells are of the leukocyte lineage (12). The purified satellite cells are viable (both in myofiber cultures and cultured alone) and give rise to myogenic cells, e.g., desmin+ve proliferating myoblasts, which form denovo myotubes as well as the obligate fibroblast-like cells after 24 h of culture (Fig. 3). We find both types of cells stain positive for Syndecan-4 and m-Cadherin (14), while fusion-competent myoblasts are Pax7+ve, strongly desmin+ve (Figs. 2 and 3), and lose CD34 expression.

This protocol begins with tissue digestion, dissociation, and sedimentation to obtain purified bulk muscle fibers, and continues with a further digestion and or dissociation to strip the satellite cells from the fibers (Fig. 1). Expect the whole procedure to take a full day. Some steps are tricky, so practice on less critical samples first and of course observing the fractions under the microscope when in doubt will be helpful. Hints on culturing the derived cells and full immunodetection protocols are also provided.

2. Materials

1. Collagenase digestion medium: Collagenase type II, 150–200 Units per milliliter (U/ml), in buffered medium (e.g., DMEM, with or without serum), buffered to pH~7 (e.g., with 10 mM HEPES) (see Note 1). Collagenase may be made as a 10 or 20× stock solution in medium or Phosphate Buffered Saline (PBS) and is stable at 4°C for days or may be frozen at −20°C.
2. Buffered medium: DMEM, Ham's F10, and RPMI have all been used successfully. We also add at least 10% serum, which better preserves the integrity of the fibers (see Fig. 1). The species of serum may be unimportant since the tissue will not remain long in this medium; horse serum is typically economical, as is using old or outdated serum in the lab. Since the procedure is typically not done in a CO_2 environment, the medium should be buffered to ~pH 7; we use 10 mM HEPES. PBS without calcium or magnesium and similar physiological saline solutions (e.g., HBSS) may also be used, but it is even better to remove non- or low-adherent cells, although fiber integrity will suffer.
3. PBS: Phosphate-buffered saline, without magnesium or calcium.
4. Serum for culture of fibers and cells: Consult the relevant literature for media and serum choices for your application. Of recent note, we have found Bovine Growth Serum (Hyclone) to preferentially support the growth of myoblasts over fibroblasts, as compared to many lots of Fetal Bovine Serum.

5. Dispase/collagenase solution: 0.5–1 U/ml Dispase (Gibco #17105-014), 40 U/ml Collagenase type II (these are the standard enzymes and concentrations for the usual "primary myoblast isolation" (15) (see Note 1), resuspended in buffered medium with serum as above). Filter dispase solutions as they often contain an insoluble precipitate which appears mildly toxic to cells if carried over into subsequent culture. Dispase and collagenase may be prepared as 10 or 20× stock solutions in medium or PBS and are stable for days at 4°C or may be frozen at –20°C. Avoid repeated freezing and thawing.
6. Falcon mesh-cap FACS tubes (e.g., #352235 for up to 5 ml digest volumes) or cell strainers (#352340 for up to 50 ml volumes), or equivalent are convenient to filter cell suspensions.
7. Histopaque or equivalent whole blood separation reagent (e.g., Sigma #10831), used as per manufacturer's recommendations.
8. Ficoll-400/buffered medium for step gradients: 30% and 20% Ficoll (w/v) in buffered medium.
9. Miltenyi CD45 Microbeads magnetic beads and MACS magnet separation kit. http://www.miltenyibiotec.com
10. Coated tissue culture plates or dishes: Collagen, laminin, fibronectin, diluted Matrigel, or Sigma ECM at 1–10 micrograms per square centimeter (μg/cm^2), all support myogenic cell attachment (see Note 10).
11. All sedimentation steps are performed using swinging bucket type centrifuge rotors.

3. Methods

3.1. Harvesting the Muscles of Interest

1. Euthanize the animal humanely and appropriately.
2. Dissect out the muscles of interest. We routinely take the mouse tibialis anterior, gastrocnemius, quadriceps, biceps femoralis ("hamstrings"), gluteus and any other major muscles of the leg and hip, the triceps in the upper arm, and the pectoralis if the mouse is big.
3. To limit nonmuscle cells, avoid harvesting tendon, nerve, major blood vessels, fat and connective tissue, and pick out any obvious nerve, fat and viscera.
4. Muscle should be harvested within about 5 min to remain viable. Dissect with the whole animal on ice if you think it will take longer.

3.2. Digestion of Muscle into Fibers

1. Place muscle as it is collected into the collagenase digestion medium at approximately 1 ml medium per 0.1 g muscle, in a suitable vessel with room enough for the muscle to slosh around in the solution.
2. Digest at 37°C with gentle agitation for 20 min to 2 h, the lower limit being appropriate for muscle from neonatal animals and the upper for adults.
3. The muscle is digested sufficiently when fat has been liberated but before the muscle falls apart or "fans" into fibers. Overdigestion leads to loss of satellite cell yield, but underdigestion only makes the dissociation step more difficult.

3.3. Washing of Muscle

1. After digestion, let the softened muscle settle for a few seconds in the tube. One may perform all the steps except digestions at room temperature if the cells will be cultured afterwards, alternatively, most steps may also be performed on ice with cold reagents to best preserve the cellular state.
2. Aspirate the floating debris, fat blobs and collagenase solution, add >20 volumes of PBS, resettle for a moment and aspirate more junk.
3. Pour the muscle into a petri dish. A 10 cm dish works well up to a gram of muscle, and a 15 cm dish up to 5 g; for larger amounts, process multiple batches.
4. Add and aspirate more PBS, removing more fat blobs, nerve strings, blood vessels, mouse fur, and any other debris. Note that some muscle chunks may begin to fall apart so do not aspirate those, and if this happens consider less digestion next time. When in doubt, look with a magnifier: muscle has fibers, but fat and connective tissue do not.

3.4. Dissociation of Muscle into Fibers

1. To the muscle add buffered medium: about 5 ml for a 10 cm dish or 10 ml for a 15 cm dish.
2. Triturate to disrupt the muscle (see Fig. 1a and Note 2). The muscle will plug the pipette at first, but by rolling the plug and broken pipette tip against the bottom of the dish, the muscle will pass through the tip.
3. Triturate until all of the muscle can pass easily through the pipette.

3.5. Horizontal Sedimentation of Fibers from Debris

1. Tilt the dish to pool the liquid, muscle chunks and fibers on one side of the dish, then tilt the dish slowly in the other direction to drain the liquid and fibers away from the larger muscle chunks and debris (see Fig. 1b).
2. Use the pipette to herd stray chunks and debris away from the pooling fibers (see Note 3).

3. Pipet the fibers into a new tube. A 15 ml conical tube will be adequate for 1 g of muscle, and a 50 ml tube will be adequate for over 2 g.

3.6. Redissociation of the Remaining Muscle

1. Repeat steps in Subheading 3.4 above. Switch to a pipette with a smaller opening; if you are using the broken plastic pipette, switch to an unbroken plastic pipette for a smaller effective bore.
2. Repeat steps in Subheading 3.5 above (see Note 4).

3.7. Sedimentation of Fibers from Interstitial Cells

1. Add buffered medium with serum or PBS to the fibers to fill the tube, and allow the fibers to settle (see Note 5 and Fig. 1c). Settle for about 10 min or until a discreet fiber layer is seen. If the sample is too viscous and no discreet pellet forms after 10 min, then sediment at 200 × *g* for 2–5 min (or at half the speed typically used to pellet cells), or dilute the sample more and resettle. We typically sediment this first round and then settle by gravity for the subsequent round(s), below.
2. Save the supernatant for Subheading 3.8 (optional) if desired, or discard if the interstitial cells are not desired. If the muscle is overdigested, many myogenic cells will be in this supernatant.
3. Add more buffered medium or PBS and resettle the fibers (see Note 6). Watch to see the different rate of settling of size fibers and contaminants: one can pipet some contaminants and fiber bundles off the bottom of the tube as they settle early, and one can aspirate the supernatant early to remove the smallest fiber fragments (see Fig. 1d). If the fibers are going to be cultured ex vivo (Fig. 3), then settle them once more through medium to avoid introducing too much PBS to the culture (see Note 6).

3.8. (Optional) Preparation of Bulk Interstitial Cells: Cells that Are Found in Muscle but Are Not Adherent to the Fibers (see Note 7)

1. Sediment the collected supernatant from step 2 in Subheading 3.7 above for 5 min at 500 × *g* and aspirate the supernatant.
2. Resuspend the pellet in PBS and sediment residual fiber fragments for 1 min at 200 × *g*.
3. Filter the supernatant through 40 or 50 μg mesh cap tubes or strainers.
4. Sediment the cells from the filtrate for 5 min 500 × *g*. The cell pellet contains mostly debris on top, assorted interstitial cells toward the bottom, and red erythrocytes packed at the very bottom.
5. Aspirate the supernatant. For less debris, tilt the tube and allow the top layer to slide and be aspirated also.

For less erythrocytes, resuspend the pellet gently and immediately so as not to dislodge the packed red blood cell pellet at the very bottom, or sediment over Histopaque as per the directions.

6. Wash again in PBS or medium and resediment to get a cleaner prep.

3.9. Preparation of Satellite Cells from Fibers (see Note 8)

1. *Variation: enzymatic digest.* This option gives a cleaner prep than *Variation 3.9.2* below, and may save hands-on time when many samples are processed at once.
   (a) Resuspend fibers in medium (~20 settled-fiber volumes) and allow the fibers to resettle.
   (b) Aspirate the supernatant and resuspend in another 10 volumes of buffered medium for digestion.
   (c) Add dispase/collagenase to the solution, and digest 40 min to an hour at 37°C with moderate agitation. When the sample is well digested you may see a stringy "ball," something like a fresh genomic DNA precipitate. Both enzymes are relatively safe to the cells and a longer digestion gives a cleaner prep.
2. *Variation: needle shred of fibers.* This is a scaled up version of (16). This Variation is very quick for a few samples and gives viable cells, but we find it too contaminated with myofibrils and other cellular debris for some types of analyses, such as flow cytometry, without additional processing. Some have reported better retention of cell surface markers and/or better cell adhesion using this variation.
   (a) Using an appropriately sized syringe, pass the fiber suspension quickly but without cavitation through a 19- or 20-gauge needle 20–25 times until the fibers are completely shredded.
3. *Variation: a combination of the above variations.* This gives the cleanest prep of the three, high yield and does not take much more time than *Variation 3.9.1*.
   (a) Pass the fiber suspension from 3.7. through a 19- or 20-gauge needle 5 or 6 times.
   (b) Add dispase/collagenase to the suspension for 30 min.
   (c) Incubate for 30 min at 37°C.
   (d) Vortex briefly or pass through the needle once or twice again.
4. From the above variations, further purify the cells from debris:
   (a) Sediment the remaining fiber debris 1 min, 200 × *g*, and save the fiber pellet on ice. (In case the digest and/or shred was not efficient, the fiber pellet can be redigested.)

(b) Some debris floats so next filter the supernatant through 40–50 μm mesh.

(c) Sediment the cells for 5 min at 500×*g* (see Note 9).

(d) Resuspend the cells with cold PBS, sediment for 5 min at 500×*g*, and remove the supernatant to remove much of the debris. Repeat, until the size of the cell pellet does not decrease substantially.

3.10. Further Optional Purification Steps

1. (Optional) Density purification of cells from debris:

   Because it may be performed in the cold, this optional method best retains the current snaphot of the cell state, as compared with further culture of the cells.

   (a) Prepare a Ficoll-400/buffered medium step gradient in a narrow tube, layering 30% and 20% Ficoll (w/v), with 0.5–1 ml each step (per approximate gram of original muscle).

   (b) Resuspend cells in 0.5 ml of buffered medium, and layer this on top of the Ficoll gradient.

   (c) Sediment in a swinging bucket rotor at 5,000×*g* (or your typical speed to pellet bacteria), for 1 h with no brake. Most of the cells will settle on top of the 30% layer and at least half of the debris will be elsewhere.

   (d) Carefully remove the top layers and pipet the cell layer into a separate tube.

   (e) Dilute tenfold or more in buffered medium and resediment 5 min at 1,000×*g* to pellet the cells.

2. (Optional) Adherence purification of cells from debris:

   (a) Plate prep onto coated dishes. "Quiescent" satellite cells will attach well enough in a couple of hours while debris will not attach for days.

   (b) Observe by microscopy every hour and after the cells attach, remove the medium with floating debris, wash with PBS as needed and harvest the cells (see Note 10).

3. (Optional) Immunoselection against leukocytes:

   Injured muscle from 2 to 5 days is typically infiltrated by immune cells. While most of these cells are weakly adherent and will detach from the bulk fibers through PBS washes, significant numbers of infiltrating cells may still be present in the final cell fraction, especially if buffered medium is used instead of PBS. As most of the immune express CD45 and the vast majority of myogenic cells do not, the CD45 cells can either be identified and gated in flow cytometry analysis, sorted on the FACS (see Chapter 11), or removed by magnetic bead sorting as follows.

(a) On ice, bind CD45 conjugated magnetic beads to the cell sample for 15 min. Use buffered medium with serum instead of the specified EDTA/saline buffer to retain cell viability.

(b) Wash as per the manufacturer's directions (but with buffered medium if viable cells are desired), and pipet into the magnetic column to bind beads.

(c) Wash as directed (with the EDTA saline), and elute into buffered medium with serum to maintain cell viability.

(d) Repeat magnetic sorting as needed. We find that the bead sorter only removes about two-thirds of the immunostained cells on the first pass, so a second sort is typically required.

4. Notes

1. US Biological, Worthington, Sigma and Gibco/BRL brands of collagenase and dispase have been used and work well. We find variation in activity among different brands and lots, and recommend testing for the degree of digestion: that is, to ensure that the muscle is still intact after 1.5 h of digest at 37°C with gentle agitation, but dissociates readily when triturated. Other types of collagenase (I, IV) may also work for your application but test that your liberated cells ultimately retain the appropriate cell surface markers. We used a modification of this method to obtain high yields of viable, albumin+ve hepatocytes and thought it simpler and easier than methods requiring perfusion of the liver. Hepatocyte researchers may want to try this: Briefly the liver was dissected on ice, minced with a razor in cold PBS/1 mM EDTA, triturated briefly through a pipette and then settled for a few minutes in cold PBS/EDTA to remove much of the blood, then digested for 1 h in the collagenase/dispase solution referenced (or see Subheading 3.5, step 1). The digest was further triturated, filtered through 50 μm mesh, and the cells pelleted by sedimentation. Plated onto coated dishes (see Note 10), the hepatocytes adhere while the remaining blood, mostly erythrocytes, can be washed off.
2. For up to a gram of muscle we use a glass Pasteur pipette that has had the tip broken off to give a ~5 mm opening (see Fig. 1a). Flaming to polish the tip will give less broken fibers, which desirable for culturing bulk fibers, but a sharp broken tip dissociates the muscle easier, which increases yield. For neonatal muscle an unbroken Pasteur pipette will suffice.

Use a rubber bulb to work the pipette, not a motorized pump. For larger amounts of muscle, try the first trituration with a 5 or 10 ml plastic pipette that has the tip segment broken off, or a 25-ml pipette which typically has a large bore opening. Triturate as above using the motorized pump and let the open bore ride the bottom of the dish while pipetting out so as to disrupt the muscle, yet not splash a mess. One may also dissociate in a tissue culture flask to contain the sample if splashing or loss of material becomes a problem.

3. The digest is likely thick at this part and in general this is a difficult step. So here are some hints: work on a dark surface or with side-light to see the light-hued fibers and lighter tendons; keep the volumes low, around 5 ml in a 10-cm dish and 10 ml in a 15-cm dish; before pooling and draining the triturate, swirl it around the dish once to coat the plastic with a lubricious film of digested collagen to achieve a laminar draining of fibers from the chunks.
4. A third or fourth redissociation will yield a few more fibers and may or may not be worth it. We typically do not redigest the remaining tissue, but one might get more fibers from it. The remaining connective tissue is rich in fibroblasts, remaining myogenic cells and may have chondrocytes or other cells of potential interest, which will grow out if cultured.
5. From this point on, the use of buffered medium with serum supports healthier fibers and cells for further culture, but PBS does a better job of dislodging the more loosely adherent interstitial and immune cells from the fibers. Also, for protein or nucleic acid analysis it would be prudent to keep all solutions, preparation, and the centrifuge cold except during the actual digestion steps.
6. These washes are to remove the smaller fiber fragments and interstitial cells, including blood cells. Extensive washes with PBS, along with overdigestion in collagenase, will also detach fiber-associated cells from injured fibers. It would be wise to save these washes (the interstitial fraction), on ice for the first few tries at this protocol, and in case the protocol gives low yields. For culturing bulk fibers ex vivo, we find that less fibers give more cells: dilute the fiber volume (loosely settled) 40-fold or more in growth medium to get good satellite cell activation and proliferation. Culturing the fiber-associated cells on the fibers gives better activation than culturing cells alone as they seem to prefer the fiber as the niche. The use of the richer DMEM-based media with chick embryo extract (derivatives of (17)) also works very well for this initial activation and proliferation of satellite cells on fibers.

7. There will be also myogenic cells in the "interstitial" fraction of this prep, the sum of the satellite cells attached to very small myofiber fragments, cells digested and detached from the fibers and those myogenic cells residing in the interstitium. Thus, this protocol is better at removing interstitial cells from the fibers than removing satellite cells from the interstitial population.
8. If you are processing cells from ex vivo cultured fibers, pour the fibers and medium into a tube and allow the fibers to settle. Also, we noticed that if the fibers are cultured in low serum medium instead of the typical 20% serum, then they tend to fall apart into myofibrils and such after digestion (Fig. 1c), which makes it more difficult to gate on the true cells by FACS, unless the debris is removed by extra washes and/or adherence purification, as described below.
9. There may be a fluffy layer of debris just above the denser, slightly sticky cell pellet, particularly from *Variation 3.9.2* or from ex-vivo cultured samples. Gently tilt the tube and aspirate this layer if you see it. The cells are now ready for culturing if a lot of debris can be tolerated. View the cells under a hemacytometer to get a cell count and assay the amount of debris. An additional wash or two with buffered medium or PBS will remove more debris.
10. Live cells will readily adhere to coated tissue culture dishes and plates, but different cell types adhere to different degrees. In fact, since fibroblasts adhere to uncoated tissue cultureware while myogenic cells do not, preplating on uncoated dishes is the easiest way to remove fibroblasts (18). Differential attachment can be used to separate cell types or cells from debris, where the myogenic cells will adhere to coated substrates within hours but debris will not. Also, we routinely do not use trypsin to detach myoblasts, but instead wash the plate two to three times with PBS, add a small volume of PBS to the dish and place it at 37°C for 2 min. The myoblasts will then detach before fibroblasts or myotubes and will better retain surface proteins. For clonal growth, the choice of serum lot, addition of basic FGF and the use of conditioned medium are essential during the initial plating and outgrowth. Test different FBS lots for their support of clonal growth and purchase those lots. We have also found Bovine Growth Serum (Hyclone) to preferentially support myogenic growth over fibroblast growth or fate, as compared with many FBS lots. Cultures are typically fed every 1–2 days. "Conditioned" medium is 1-day-used medium from a healthy myoblast culture, sterile-filtered and mixed 1:1 with fresh medium.

Acknowledgements

We would like to thank members of the Conboy and Wagers laboratories for constructive feedback on these protocols, John Wang for assistance with Fig. 2, and Matthew O'Connor for tips on magnetic bead sorting. We thank the National Institutes of Health, National Institute on Aging (R01 AG027252), the Stem Cell Research Foundation/American Health Assistance Foundation, California Institute for Regenerative Medicine and Ellison's Medical Foundation support to I.M.C.

References

1. Collins CA, Olsen I, Zammit PS, Heslop L, Petrie A, Partridge TA, Morgan JE (2005) Cell 122:289–301
2. Irintchev A, Zeschnigk M, Starzinski-Powitz A, Wernig A (1994) Dev Dyn 199:326–337
3. Beauchamp JR, Heslop L, Yu DS, Tajbakhsh S, Kelly RG, Wernig A, Buckingham ME, Partridge TA, Zammit PS (2000) J Cell Biol 151:1221–1234
4. Conboy IM, Rando TA (2002) Dev Cell 3:397–409
5. Carlson ME, Hsu M, Conboy IM (2008) Nature 454:528–532
6. Conboy IM, Conboy MJ, Smythe GM, Rando TA (2003) Science 302:1575–1577
7. Sherwood RI, Christensen JL, Conboy IM, Conboy MJ, Rando TA, Weissman IL, Wagers AJ (2004) Cell 119:543–554
8. Mitchell PO, Mills T, O'Connor RS, Kline ER, Graubert T, Dzierzak E, Pavlath GK (2005) Dev Biol 283:240–252
9. Conboy MJ, Karasov AO, Rando TA (2007) PLoS Biol 5:e102
10. Olguin HC, Olwin BB (2004) Dev Biol 275:375–388
11. Clark WE (1946) J Anat 80:24–36
12. Brack AS, Conboy IM, Conboy MJ, Shen J, Rando TA (2008) Cell Stem Cell 2:50–59
13. Bischoff R (1975) Anat Rec 182:215–235
14. Wagers AJ, Conboy IM (2005) Cell 122: 659–667
15. Rando TA, Blau HM (1994) J Cell Biol 125:1275–1287
16. Shefer G, Wleklinski-Lee M, Yablonka-Reuveni Z (2004) J Cell Sci 117: 5393–5404
17. Yaffe D, Fuchs S (1967) Dev Biol 15: 33–50
18. Richler C, Yaffe D (1970) Dev Biol 23: 1–22

Chapter 11

Immuno-Analysis and FACS Sorting of Adult Muscle Fiber-Associated Stem/Precursor Cells

Michael J. Conboy, Massimiliano Cerletti, Amy J. Wagers, and Irina M. Conboy

Abstract

We describe the immunostaining methods we commonly use to detect the more robust cell markers identifying the various cell populations obtained by the enzymatic and or mechanical dissociation of muscle satellite cells: CD34, m-Cadherin, and Pax7, self-renewing muscle stem cells expressing CXCR4 and β1-integrin, populations of proliferative myogenic progenitor cells expressing, Pax3, Pax7, Myf5, MyoD1, and desmin, differentiating myoblasts expressing myogenin and eMHC, and the CD45 expressing leukocyte lineage cells that infiltrate injured and regenerating skeletal muscle.

Key words: Satellite cell, Myoblast, Myogenic, Renewal

1. Introduction

A highly enriched but mixed myogenic cell population is obtained by enzymatic or mechanical dissociation (Chapter 10, and refs. (1, 2)). These cells can be further purified by fluorescence-activated cell sorting (FACS) based on a panel of cell surface markers (CD45-Sca1-Mac1-CXCR4+β1-integrin+, abbreviated CSM4B) that identify a self-renewing, transplantable muscle stem/precursor cell population (3, 4). Sorted CSM4B skeletal muscle precursor cells (SMPs) represent a subset of canonically defined Pax7+ satellite cells, and when transplanted, contribute robustly to in vivo muscle regeneration (contributing up to 94% of myofibers and significantly improving muscle contractile function in dystrophic recipient mice). Importantly, transplanted SMPs also reengraft and renew within the satellite cell compartment, seeding the endogenous pool of myogenic precursors and participating in subsequent

Irina M. Conboy et al. (eds.), *Protocols for Adult Stem Cells*, Methods in Molecular Biology, vol. 621,
DOI 10.1007/978-1-60761-063-2_11, © Springer Science+Business Media, LLC 2010

rounds of injury repair. This chapter also details our general immunostaining protocols for analysis of myofibers with attached cells, processed and enriched cells or sorted cells. For more background information on muscle derived cells, see Chapter 10.

2. Materials

2.1. Immunostaining Reagents

1. Antibodies: For each antibody to a target antigen, also select an isotype matched immunoglobulin (e.g., IgG2a, IgG1, IgG1kappa, and IgM), to use as a control. A control of secondary antibody only is also valuable, but should not be a replacement for the isotype matched primary plus secondary antibody control. The reason for this is that some immunoglobulins, be they primary or secondary antibody, will nonspecifically bind to certain cellular structures, typically the nucleus or basal lamina. Isotype matched control antibodies are available from the better suppliers of primary antibodies. If cell surface or lamina background staining becomes an issue, try using F(ab') fragment antibodies.
2. Staining buffer (SB): 1% or more of serum or albumin in PBS. For antibodies giving high background, use serum from the species in which the detection antibodies (e.g., dye-conjugated) are made and/or add competitor immunoglobulins from this species to the primary antibody incubation at 10 μg/ml.
3. Permeablization buffer: PBS (or SB) with 0.25% Triton-X 100.
4. Ethanol fixative: 70% ethanol/30% water, or 75% ethanol/25% glacial acetic acid. Use 100% or 200 proof ethanol, not most "denatured" types (although small amounts of methanol or isopropanol are acceptable). The fixative with acetic acid better preserves nuclear morphology, but may kill your antigen (such as Pax7, myoD), so test this.
5. Paraformaldehyde, 2–4%, dissolved in hot PBS, then cooled, aliquoted, and stored at −20°C. Paraformaldehyde and its fumes are both toxic and noxious, so handle accordingly: weigh and mix in a fume hood and use personal protection. We have also tried paraformaldehyde fixative with 2% sucrose, but qualitatively find it no better or worse.
6. Citrate/Tween buffer: 10 mM sodium citrate, pH ~6.5 to 7 with 0.05% Tween-20.
7. DNase I at the higher stock concentrations (e.g., 100 U/μl.).
8. DNase digestion buffer: any physiological buffer with magnesium should work; out of convenience we use a 1:1 mixture of cell culture medium (e.g., F-10 or Minimal Essential Medium), and PBS.

2.2. Antibodies Used for Flow Cytometry/FACS Purification

1. 30-F11 (1:200) (antimouse CD45, phycoerythrin (PE) conjugate (eBioscience #12-0451-83, San Diego, CA)).
2. D7 (1:800) (antiSca-1, Ly-6A/E, APC conjugate (eBioscience #17-5981-82)).
3. M1/70 (1:200) (antimouse CD11b, PE conjugate, (eBioscience #12-0112-82)).
4. Primary: CXCR4 (1:100) (biotinylated antimouse CD184 (BD Pharmingen #551968)); Secondary: Streptavidin (1:100) (Cy7-PE conjugate (eBioscience #25-4317-82)).
5. Primary: β1-Integrin (1:200) (antimouse CD29, purified, (BD Pharmingen #553837, San Jose, CA)); Secondary: AH (1:100) (antiarmenian hamster IgG, fluorescein isothiocyanate (FITC) conjugate (eBioscience #11-4111-85)).

2.3. Antibodies for Staining of Myofiber-Associated Cells for SMP Sorting from GFP Transgenic Mice

1. 30-F11 (1:1,600) (antimouse CD45, allophycocyanin (APC) conjugate (BD Pharmingen #559864)).
2. D7 (1:800) (antiSca-1, Ly-6A/E, APC conjugate (eBioscience #17-5981-82)).
3. M1/70 (1:800) (antimouse CD11b, APC conjugate, (eBioscience #17-0112-82)).
4. Primary: CXCR4 (1:100) (biotinylated antimouse CD184 (BD Pharmingen #551968)); secondary: streptavidin (1:100) (Cy7-PE conjugate (eBioscience #25-4317-82)).
5. β1-integrin (1:400) (antimouse/rat CD29, PE conjugate, (Biolegend #102207, San Diego, CA)).

3. Methods

3.1. Immunodetection of Antigens

1. Detection of surface antigens (e.g., CD34, m-cadherin, Sca1, CXCR4, and CD45), on live cells, e.g., for FACS sorting. With live cells, all reagents must be cold and incubations done on ice (see Notes 1 and 2).
   (a) Block, if needed, with IgG or serum in staining buffer for 15 min on ice. To block nonspecific binding, pick an IgG or serum from the species in which the secondary antibody was made.
   (b) Add primary antibody in SB; incubate for 15 min.
   (c) Wash three times with SB.
   (d) Add secondary antibody in SB, including any dyes such as live Hoechst; incubate for 15 min.
   (e) Wash three times with SB.

(f) Analyze by FACS or microscopy. As noted, live cells will begin to endocytose the antigen/antibody complex when warmed above 10–15°C, so store on ice.

2. Detection of antigens preserved by ethanol fixation (e.g., Pax3 Pax7, BrdU, desmin, eMHC, nuclear Notch1).

(a) Fix the cells in ice-cold ethanol (either 70% ethanol or 75%/25% ethanol/glacial acetic acid) for at least 15 min. The samples may be stored at this point in fixative for months at −20°C.

(b) Remove the ethanol and rehydrate sequentially by washing with cold 50% ethanol/PBS, then cold 25% ethanol/PBS, then PBS alone. Alternatively, one may slowly mix in cold PBS to dilute the fixative.

(c) ADDITIONAL steps for detecting bromo-deoxyuridine (BrdU) and analogs:

- Remove PBS; incubate with fresh 2.5 M HCl for 30 min at room temperature (RT) to nick the nucleic acids so as to be accessible to the antibody.
- Wash briefly three times with PBS to remove the acid.

(d) Permeabilize with 0.25% Triton-X 100 in PBS for 5 min at RT (for cells) or 10 min at RT (for fibers).

(e) Block, if needed, with IgG or serum in SB.

(f) Immunolabeling:

- Add primary antibody in SB; incubate for at least 4 h at RT or overnight at 4°C.
- Wash with PBS (short wash, 1 min at RT).
- Wash with permeablization solution (above) for 1 min at RT.
- Wash one more time with SB (this last wash should go for 30 min, to leach out residual unbound antibody from the nucleus and to reblock).
- Add secondary antibody in SB; incubate for >4 h at RT or overnight at 4°C.
- Wash 3× with PBS. The first two washes are short, 1 min at RT, the last is 30 min).

(g) Analyze by FACS or add antifade reagent and mount for microscopic examination.

3. Detection of antigens preserved by paraformaldehyde (PFA) fixation (e.g., for cell surface antigens, desmin, pax3, pax7, and BrdU).

(a) Fix in cold 2–4% PFA on ice for 5 min (cells) or 10 min (fibers) (see Note 3).

(b) Wash three times in PBS (the last wash should be for at least 1 h). The samples may be stored in PBS for up to 2 weeks at 4°C.

(c) Wash once again in PBS for 1 min at RT.

(d) For surface antigens, immunolabel as per 3.1.1, above, but the reagents and cells can be kept at room temperature (see Note 4).

- Add antifade reagent and mount – OR – proceed with further staining, as follows:
- Refix at this point in 2–4% PFA on ice for 5 min (cells) or 10 min (fibers). This will retain the immunolabeled surface antigen/antibody complex through the permeablization steps (see Note 5).
- Wash three times with PBS as in step 2 above.

(e) For intracellular or nuclear antigens, permeabilize with 0.25% Triton-X 100 in PBS for 5 min at RT (for cells) or 10 min at RT (for fibers).

(f) For BrdU staining, it is necessary to partially digest the DNA with DNase as follows:

- Incubate samples for 40 min in 0.2 Kunitz units per microliter DNase I in a buffer with magnesium. We typically use PBS/MEM at 1:1, and add 100 μl buffer per 10^5 cells (see Note 6).
- Wash twice with PBS and proceed to step 3.1.3.h.

(g) (Alternative) Poaching method to denature nuclear antigens (including BrdU).

- Beginning with a PFA fixed sample that has already been washed at least once in PBS, poach 20–40 min (a double-boiler works well) in Citrate/tween buffer.
- Remove samples from the hot buffer and proceed with the immunolabeling procedure below. The samples are already permeabillized.

(h) Immunolabeling:

- Add primary antibody in SB; incubate for at least 4 h at RT or overnight at 4°C.
- Wash with PBS (short wash, 1 min at RT).
- Wash with permeablization solution for 1 min at RT.
- Wash one more time with SB (this last wash should go for 30 min, to leach out residual unbound antibody from the nucleus).
- Add secondary antibody in SB; incubate for at least 4 h at RT or overnight at 4°C.

– Wash 3× with PBS (the first two washes are short, 1 min at RT, and the last is for 30 min).

(i) Analyze by FACS or add antifade reagent and mount for microscopic examination.

3.2. FACS Sorting of Cells, General Notes

Typically cells are prepurified by one of the methods in Chapter 10, 3.10 before FACS analysis or sorting. However, with skillful gating on scatter, one can analyze the cruder output from step Chapter 10 3.9.1 or 3.9.3. Satellite cells are relatively small and nongranular by scatter (Fig. 1). Live cells are also relatively more fragile than the average lymphocyte, requiring the larger bore

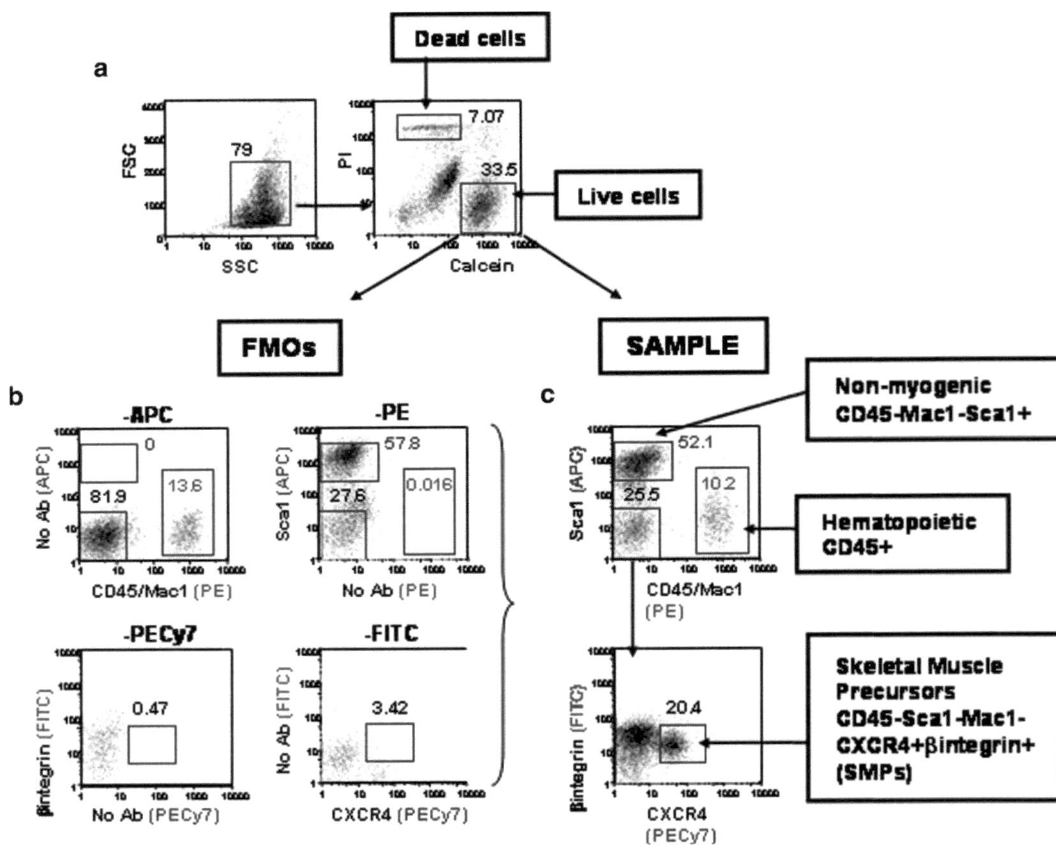

Fig. 1. Sorting profile for the purification of Skeletal Muscle Precursor Cells (SMPs). After isolation, myofiber-associated (MFA) cells are stained with antiCD45, antiSca1, antiMac1, antiCXCR4, and antiβ1-integrin antibodies, and then analyzed by flow cytometry, as in (3). (**a**) MFA are gated for size and granularity (forward and side scatter, respectively), and live cells (calcein^{+} and propidium iodide^{-}). (**b**) Fluorescence Minus One (FMO) controls, depicted in the plots, are MFA cells stained with all antibodies except one, as indicated. FMO controls are used for setting appropriate gates discriminating cell populations that stain positive or negative for a given marker. (**c**) Fluorescence activated cell sorting (FACS) using five cell surface markers (antiCD45, antiSca1, antiMac1, antiCXCR4, and antiβ1-integrin) enables separation of distinct cell populations within the MFA preparation, including nonmyogenic (CD45^{-}Mac1^{-}Sca1^{+}) cells, hematopoietic (CD45^{+}) cells, and skeletal muscle precursor (SMP) cells (CD45^{-}Sca1^{-}Mac1^{-}CXCR4^{+}β1-integrin^{+}) cells

FACS nozzle and/or lower flow pressures. Sort into buffered growth medium to maintain cell viability.

Mechanically enriched populations of myogenic cells can also be further purified by FACS based on cell surface antigens. After isolation as described above, myofiber-associated cells can be stained as described below for purification by FACS of the self-renewing, highly myogenic SMP population. Markers for SMP isolation are CD45$^-$Sca-1$^-$Mac-1$^-$CXCR4$^+$β1-integrin$^+$ (CSM4B), as described in (3).

3.3. Staining of Myofiber-Associated Cells for SMP Sorting

1. For antibody staining, resuspend cells at indicated concentrations of antibody and incubate 20 min on ice.
2. After incubation, fill the tube with HBSS/2% FBS, centrifuge at 4°C at 500×g for 5 min (SM = Hank's Buffered Saline Solution (HBSS, (Gibco)) + 2% donor bovine calf serum). Aspirate supernatant from cell pellet and proceed to secondary antibody staining.
3. Resuspend cells in secondary antibody mix. Incubate 20 min on ice. Fill the tube with HBSS/2% FBS, centrifuge at 4°C at 500×g for 5 min.
4. Remove the supernatant. Resuspend in 300 μl HBSS/2% FBS, add propidium iodide (PI, 1:1,000 dilution of a 1 mg/ml stock), Calcein blue (1:1,000 dilution of a 4.65 mg/ml stock, Molecular Probes #C1429).

3.4. (Alternative Protocol) Staining of Myofiber-Associated Cells for SMP Sorting from GFP Transgenic Mice

1. Resuspend cells at indicated concentrations of antibody and incubate 20 min on ice (see Subheading 2.3).
2. Fill the tube with HBSS/2% FBS, centrifuge at 4°C at 1,200 rpm for 5 min (SM = Hank's Buffered Saline Solution (HBSS, (Gibco)) + 2% donor bovine calf serum).
3. Incubate 20 min on ice. Fill the tube with HBSS/2% FBS, centrifuge at 4°C at 1,200 rpm for 5 min.
4. Remove the supernatant. Resuspend in 300 μl HBSS/2% FBS, add PI (1:1,000), and Calcein (1:1,000).

3.5. FACS of SMPs

Live cells are identified by positive staining with calcein blue and negative staining for propidium iodide (PI) (Fig. 1a). Use fluorescence minus one (FMO) controls, in which multiple tubes of control cells are stained with all antibodies except one, to determine appropriate placement of sort gates (Fig. 1b). SMPs are sorted as CD45$^-$Sca-1$^-$Mac-1$^-$CXCR4$^+$β1-integrin$^+$ (CSM4B) (Fig. 1c), as described in (3), and should be double-sorted to ensure purity. We typically obtain ~50,000 SMPs per gram of muscle from uninjured, 8–12-week-old C57/BL mice. Flow cytometry analysis and cell sorting are performed using high-speed, multiparameter cell sorters. Machines must be optimized for flow rate and nozzle size to achieve maximal cell purity and viability. We use either a BD FACSAria (70 μm nozzle) or Cytomation Moflo (60 μm nozzle). Flow cytometry data can be analyzed offline using FlowJo software (Tree Star, Inc., Macintosh version 8.1.1, Ashland, OR).

3.6. Clonal SMP Cell Culture

SMPs can be cultured in bulk or clonally on collagen/laminin-coated plastic in F10 containing 20% horse serum and 5 ng/ml bFGF. Single cell cloning assays are performed in 96 well plates, with fresh bFGF (5 ng/ml, Sigma, St. Louis, MO) added daily. Clones may be replated on day 5, and switched on day 7 into myogenic differentiation medium (DMEM (Gibco) containing 1% FBS and lacking bFGF). Myogenic differentiation is assessed by staining with differentiation markers (desmin, MyHC) and by the observation of fused, twitching myotubes by brightfield microscopy.

4. Notes

1. When staining live cells for surface antigens one must inhibit endocytosis of the antigen and/or antibody. This is accomplished by using cooled reagents and incubating the cells on ice. For all immunostaining, the use of isotype matched antibodies as a negative control for the primary antibody is essential, and given the number of steps in these protocols and the variety of cells obtained, such controls are worth the time and should be included in all experiments. If one uses only the secondary antibody as the only negative control, then one may miss nonspecific binding of the primary immunoglobulin, and witness a false-positive signal. The use of an additional secondary antibody only control is also useful to identify nonspecific secondary binding. We typically use starting antibody concentrations of around 1 μg/ml for monoclonal antibodies, 10 μg/ml for polyclonal purified antibodies, and a 1:10 dilution for unpurified antiserum. If much more antibody needs to be used, then a different antibody or detection method should be considered.
2. For handling cells attached to slides, we typically just flick the liquid off into the sink and pipet on fresh liquid. This avoids touching/scraping any part where the cells may be attached to the slide. Volumes of around 100 μl/cm^2 are typically used for antibody incubations, with at least double for washes. For cells in tubes, such as the common 5 ml FACS tubes, if the tube is inverted briefly, immediately after centrifugation the liquid will be decanted, yet the cell pellet will faithfully remain. Do not tap the tube to remove the last drop and do not let it

sit inverted to drain. Using a vacuum aspirator induces a liquid flow in the tube that will typically suck up the cell pellet. Flick the tube gently (while upright) to disrupt the cell pellet in the remaining residual liquid before adding the next buffer. Volumes of around 100 μl per 10^6 cells are typically used, and washes of 2 ml per 5 ml sized tube.

3. Paraformaldehyde fixation adds a certain amount of background to immunostaining. We have given a minimal fixation which gives a minimal background that does not seem to be improved by treatments such as glycine or borate. If more fixation is needed or the background is still too high, explore the literature for methods of reducing such background.
4. For staining many surface antigens, do not permeabilize! If you also need to stain for intracellular antigens, first immunostain for the surface antigens without permeabilization, refix, and then permeabilize for the intracellular staining.
5. We have found the Alexa antibodies and R-phycoerythrin to be resistant to this mild fixation. FITC does not seem to be as resistant.
6. An optimal level of DNA degradation is achieved when the nuclear fluorescence of a DNA stain, such as Hoechst, DAPI or propidium iodide, has been reduced by about half. The researcher would be wise to test DNase digestion time and enzyme concentration on test samples before the crucial ones.

Acknowledgments

We thank the National Institutes of Health, National Institute on Aging (R01 AG027252 and K01 AR050515 to I.M.C.), the Stem Cell Research Foundation (Maryland, to I.M.C.), and the Harvard Stem Cell Institute (Cambridge, MA, to A.J.W.) for recent and current funding of these endeavors.

References

1. Conboy IM, Conboy MJ, Smythe GM, Rando TA (2003) Notch-mediated restoration of regenerative potential to aged muscle. Science 302:1575–1577
2. Collins CA, Olsen I, Zammit PS, Heslop L, Petrie A, Partridge TA, Morgan JE (2005) Stem cell function, self-renewal, and behavioral heterogeneity of cells from the adult muscle satellite cell niche. Cell 122: 289–301
3. Sherwood RI, Christensen JL, Conboy IM, Conboy MJ, Rando TA, Weissman IL, Wagers AJ (2004) Isolation of adult mouse myogenic progenitors: functional heterogeneity of cells within and engrafting skeletal muscle. Cell 119:543–554
4. Cerletti M, Jurga S, Witczak CA, Hirshman MF, Shadrach JL, Goodyear LJ, Wagers AJ (2008) Highly efficient, functional engraftment of skeletal muscle stem cells in dystrophic muscles. Cell 134:37–47

Chapter 12

Altered Expression of NFκB in Ex Vivo Differentiated Dendritic Cells from the Aged Subjects: Implications in Immunotherapy

Anshu Agrawal

Abstract

Dendritic cells are the key initiators and organizers of the immune response. They are specialized antigen-presenting cells whose immunogenicity leads to the induction of antigen-specific immune responses. They are therefore ideal vehicles to deliver strong immunogenic signals to eliminate tumors. Dendritic cells can dictate the polarization of T cell responses by virtue of the nature of their cytokine secretion. They are also critical for maintenance of peripheral self-tolerance. Because of their key role in these central processes, dendritic cells are increasingly being used for immunotherapy of various diseases. This has been greatly aided by the development of methods to derive dendritic cells ex vivo from human adult stem cells and other progenitors. Here we describe methods to assay the activity of transcription factor NFκB in the ex vivo derived dendritic cells and discuss the probable effects of the altered expression of NFκB on the functions of these dendritic cells.

Key words: Dendritic cells, NFκB, Flow cytometry, Transcription factor ELISA

1. Introduction

Dendritic cells (DCs) are professional antigen presenting cells, which possess an extraordinary capacity to stimulate naive T cells and initiate primary immune responses (1). This established function of DCs along with methodologies to derive dendritic cells in vitro from human adult stem cells and monocytes has now offered the hope to apply DC-based immunotherapy for cancers (2). DCs have been used in numerous clinical trials to induce antitumor immune responses in cancer patients. The studies carried out to date have demonstrated that DCs pulsed with tumor antigens can be safely administered, and this approach produces antigen-specific immune responses. Recent studies also

Irina M. Conboy et al. (eds.), *Protocols for Adult Stem Cells*, Methods in Molecular Biology, vol. 621,
DOI 10.1007/978-1-60761-063-2_12, © Springer Science+Business Media, LLC 2010

suggest that DCs play critical roles in the induction of peripheral immunological tolerance, regulate the types of T cell immune responses, and function as effector cells in innate immunity against microbes (3). FLT-3 (Fms like tyrosine kinase -3) ligand, and to a lesser extent GM-CSF (granulocyte macrophage colony stimulating factor), may represent the key DC growth and differentiation factors in vivo. DCs can also easily be generated ex vivo from peripheral blood monocytes or bone marrow/circulating hematopoietic stem cells cultured in the presence of cytokine cocktails. CD34+ progenitors in the blood can be differentiated into DCs using a combination of GM-CSF and TNF-α. In the elderly, where the numbers of CD34+ progenitors are limited, monocytes can be used as precursors for DC differentiation in place of CD34+ cells.

Members of the nuclear factor kappa B (NFκB) family of transcription factors p50, p52, p65 (RelA), c-Rel, and RelB are expressed at relatively high levels in dendritic cells (DC). To varying degrees, the NFκB transcription factors are associated with dendritic cell development, maturation, and function (4–7). Altered expression of NFκB transcription factors either in a diseased state or as a function of age has a profound effect on the dendritic cell functions (5). The predominant form of inducible NFκB in antigen presenting cells such as dendritic cells is considered to be the p50/p65 heterodimer. However, p50 subunit cannot initiate transcription since it does not contain a DNA binding sequence. Therefore, measure of activation of p65 is considered an adequate measure of NFκB activity.

We have shown that DCs from aged subjects express high basal level of NFκB activation as evidenced by the high expression and phosphorylation of the p65 subunit (Fig. 1). NFκB transcription

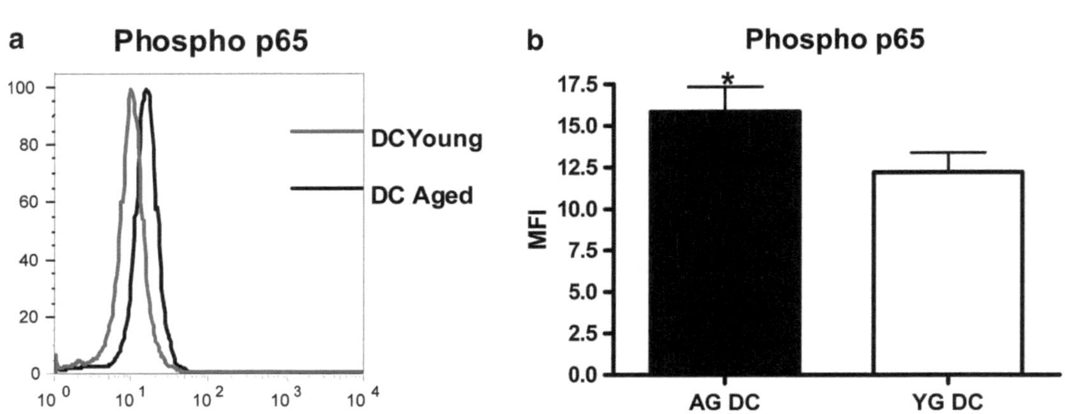

Fig. 1. (**a**) Basal level of phospho 65 unit of NFκB in DCs from aged and young as determined by flow cytometry (**b**) MFI-Mean Fluorescence Intensity of phospho p65 in DCs from aged and young. Mean± S.E. of 9 experiments

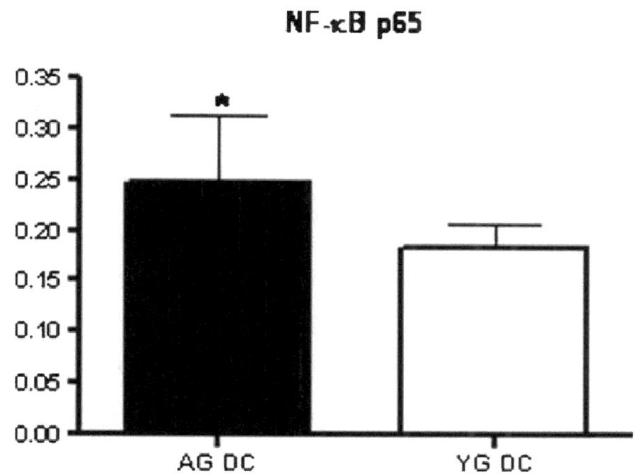

Fig. 2. Activation of p65 determined by transcription factor ELISA from Active Motif. AG DC-DCs from aged subjects and YG DC-DCs from young subjects. Mean± S.E. of 6 experiments

factor ELISA results correlate with the flow cytometry results (Fig. 2). Increased basal level of NFκB activation suggests that DCs from aged subjects exist in semiactivated state. As DCs become activated, their capacity to phagocytose antigen decreases while their antigen presenting capacity increases. Our earlier studies had shown that DCs from aged are impaired in their capacity to phagocytose antigens (8). They also secrete higher level of proinflammatory cytokines (8, 9). These altered properties of DCs need to be taken into account when using DCs in the elderly for immunotherapy of cancer. Increased secretion of proinflammatory cytokines by DCs from the elderly has a major effect on tolerance. The semiactivated state of DCs results in loss of peripheral self-tolerance through increased reactivity to self-antigens (9). Increased secretion of proinflammatory cytokines by DCs from the elderly in response foreign antigens may also affect the polarization of T cell responses towards a more proinflammatory phenotype.

2. Materials

2.1. Generation of Dendritic Cells from Monocytes

1. 30 ml Blood from healthy donors.
2. Lymphocyte Separation medium (Invitrogen).
3. Rosewell Park Memorial Institute 1640 (RPMI 1640) (Invitrogen) medium supplemented with 10% FBS (Cellgro), 1 mM glutamine (100× solution, Invitrogen), 100 U/ml penicillin, 100 μg/ml streptomycin (Invitrogen).

4. Recombinant granulocyte macrophage colony stimulating factor (rGMCSF, Peprotech, NJ) 1 mg is reconstituted at 100 μg/ml in PBS (phosphate buffered saline). Working solutions (1 ml) can be stored at 4°C for 1 month. Rest should be stored in 1 ml aliquots at −80°C.
5. Recombinant interleukin-4 (rIL-4, Peprotech, NJ) 10 μg is dissolved at 10 μg/ml in PBS. Working solutions can be stored at 4°C for 1 month.
6. CD11c-APC antibody (BD Pharmingen).
7. HLADR-PerCP (BD Pharmingen).
8. CD14 PE (BD Pharmingen).

2.2. NFκB by Flow Cytometry

1. 4% Paraformaldehyde in PBS.
2. 90% Methanol in PBS. Store at −20°C.
3. PBS with 2% FBS.
4. Phospho-p65-PE antibody (BD Biosciences).
5. p65-PE nonphosphorylated.
6. CD11c-APC.
7. HLADR-PerCP.
8. Mouse IgG1PE isotype control antibody.

2.3. NFκB by Transcription Factor ELISA

1. Nuclear fraction extraction kit (Active Motif).
2. Bradford protein reagent (BioRad).
3. TransAM NFκB transcription factor ELISA (Active Motif).

3. Methods

The determination of activity of NFκB in DCs using methods such as western blot and EMSA are hampered by the relatively small number of DCs that can be obtained from the blood of an individual. This is especially relevant when studying patient populations where the sample size is limited. The use of flow cytometry and transcription factor ELISA to determine NFκB activation overcomes these limitations, and together the two techniques provide a comprehensive picture of NFκB activity in DCs. DCs express high levels of active NFκB allowing modification of existing protocols to utilize less number of cells to determine the activity of this signaling molecule. The use of flow cytometry to determine the activation of signaling molecules has the advantage of not requiring a purified cell population. Furthermore, the advent of phospho antibodies allows staining of the activated NFκB molecule, thus providing an efficient measure of its activity. There is only one drawback, it does not allow us to determine if the NFκB has translocated to the nucleus to bind to the DNA after phosphorylation. This can be accomplished through transcription factor ELISA. Nuclear translocation of transcription factors can be quantified in DC nuclear extracts using TransAM enzyme-linked immunosorbent assay (ELISA)-based kit (Active Motif). Nuclear extracts are plated on 96-well plates coated with the immobilized oligonucleotide containing the activated NFκB consensus site (5′-GGGACTTTCC-3′). Anti-p65 antibody is added and binding is detected with a secondary HRP-conjugated antibody and developed with TMB substrate.

3.1. Generation of Dendritic Cells from Monocytes

1. This modification of the method is first described by Dr. Steiman's laboratory (1, 10) and subsequently repeated by others (11). Dilute blood 1:1 with HBSS (Hanks' Balanced Salt Solution).
2. Add 15 ml of Lymphocyte Separation Medium in a 50 ml tube.
3. Slowly layer 30 ml of diluted blood over the Lymphocyte Separation Medium. Take care to keep the two layers separate.
4. Centrifuge at 400 × *g* for 15 min at room temperature. Centrifuge brakes should be in the off position.
5. Peripheral Blood Mononuclear Cells (PBMC) will form a layer at the interface of Lymphocyte Separation Medium and plasma while red blood cells and granulocytes will settle at the bottom of the tube. Collect the PBMC layer without touching the bottom of the tube.
6. Add 30 ml of HBSS and wash by centrifuging at 260 × *g* for 10 min.
7. Discard supernatant. Wash once more with 30 ml HBSS; however, this time spin the cells at 100 g to remove platelets.
8. Resuspend the pellet in 10 ml of RPMI 1640 containing 10% FBS.
9. Add 3 ml/well in 6-well tissue culture plate.
10. Incubate for 1.5–2 h at 37°C to allow the monocytes to adhere to the plate.
11. Collect the nonadherent fraction by swirling the plate.
12. Wash the well with another 3 ml of medium to remove the remaining nonadherent cells.

13. Add 3 ml/well RPMI 1640, 10% FBS containing 50 ng/ml GM-CSF, and 10 ng/ml IL-4.
14. Incubate for 6 days, replenishing the GM-CSF and IL-4 on the third day. Add 500 μl of the medium/well containing 50 ng/ml GM-CSF, and 10 ng/ml IL-4.
15. Six days later monocytes differentiate into myeloid dendritic cells. These cells are nonadherent. A number of clumps should be visible under the microscope.
16. Gently pipet up and down to collect the loosely adherent DCs.

3.2. Characterization of Dendritic Cells

1. Centrifuge the DCs at 160 g for 10 min and collect the cell pellet.
2. Take 1×10^5 cells in 50 μl PBS +2% PBS in a FACS tube.
3. Add 5 μl of CD11c APC and 5 μl of HLA-DR-PerCP antibodies to the tube.
4. Incubate in the dark at room temperature for 20 min.
5. Wash with 1 ml of PBS, 2% FBS by centrifuging at $160 \times g$.
6. Discard supernatant. Add 200 μl of PBS 2% FBS.
7. Acquire on FACSCalibur using the CELLQuest software.
8. Analyze results using FlowJo software.
9. Dendritic cells are CD11c APC and HLADR positive. They are CD14 PE negative.
10. Almost 90% of the cells collected are DCs at this stage.

3.3. NFκB by Flow Cytometry

1. Take 6×10^5 dendritic cells in 250 μl of PBS 2% FBS.
2. Fix the cells by adding 250 μl 4% paraformaldehyde. Incubate at 37°C for 10 min.
3. Centrifuge and collect cell pellet.
4. Resuspend pellet in 500 μl of ice cold 90% methanol and incubate for 30 min on ice. Cells can be stored in 90% methanol at –20°C for 2–3 days.
5. Centrifuge to remove methanol and wash once with PBS 2% FBS.
6. Resuspend pellet in 100 μl PBS 2% FBS. Divide it into three tubes of 30 μl each. Add 10 μl each of CD11c APC and Phospho p65-PE antibodies to tube 1, 10 μl each of CD11c APC and p65-PE antibodies to tube 3, and 10 μl each of CD11c APC and isotype-PE antibodies to tube 3 and incubate at room temperature for 30 min.
7. Wash once with PBS 2% FBS.
8. Add 200 μl of PBS and acquire the sample on FACS Calibur.

9. Gate on CD11c positive cells and determine the expression of NFκB. It should be above that of isotype (Fig. 1a).
10. Calculate MFI (Fig. 1b).
11. Use nonphosphorylated p65 antibody to determine the expression level of total p65 in DCs.

3.4. Preparation of Nuclear Extract

1. A minimum of 5×10^5 cells are required for this assay. Prepare nuclear extract using the nuclear extract kit. A brief modified procedure is described below.
2. Resuspend pellet 500 μl of ice cold PBS containing phosphatase inhibitors (nuclear extract kit).
3. Centrifuge cell suspension for 5 min at 500 rpm in a centrifuge precooled at 4°C.
4. Discard supernatant. Keep cell pellet on ice.
5. Gently resuspend cells in 500 μl 1× Hypotonic Buffer by pipetting up and down several times. Incubate for 15 min on ice.
6. Add 25 μl detergent and vortex 10 s at highest setting.
7. Centrifuge suspension for 30 s at 14,000 × *g* in a microcentrifuge precooled at 4°C.
8. Transfer supernatant (cytoplasmic fraction) into a prechilled microcentrifuge tube. Store the supernatant at −80°C until ready to use. Use the pellet for nuclear fraction collection.
9. Resuspend nuclear pellet in 50 μl Complete Lysis Buffer by pipetting up and down. Vortex 10 s at highest setting.
10. Incubate suspension for 30 min on ice on a rocking platform set at 150 rpm.
11. Vortex 30 s at highest setting. Centrifuge for 10 min at 14,000 × *g* in a microcentrifuge precooled at 4°C. Transfer supernatant (nuclear fraction) into a prechilled microcentrifuge tube.
12. Store at −80°C. Avoid freeze/thaw cycles.

3.5. NFκB by Transcription Factor ELISA

1. Thaw the nuclear extract prepared above.
2. Estimate protein using 5 μl of the extract. Add 95 μl of 1:4 diluted Bradford protein reagent in a 96 well plate. Add 5 μl of the sample. The presence of certain detergents may interfere with the Bradford or BCA assay, thus use the Complete Lysis Buffer as the blank.
3. Incubate for 10 min.
4. Read at 595 nm in an absorbance plate reader.
5. Use the rest of the extract for the transcription factor assay.
6. Follow the manufacturer's protocol for the assay.
7. To determine the amount of active NFκB in dendritic cells, normalize absorbance obtained for NFκB to protein (Divide

the absorbance obtained by transcription factor ELISA by the absorbance obtained for protein of the same sample).

4. Notes

1. Monocytes should not be allowed to adhere for longer than 2 h.
2. Reconstitution of recombinant GM-CSF and IL-4 should be done cell culture grade PBS, free of endotoxin.
3. Storing the cells in 90% methanol for 2–3 days reduces the recovery of cells.
4. Flow cytometry protocol works well with other signaling molecules such as p38, ERK and AKT kinase in DCs (7, 12). These molecules can be stained simultaneously with NFκB.
5. It can be adapted for other cells.
6. It can also be used to measure NFκB activity after activation of DCs with various ligands. Time point for those will need to be determined for each experiment.
7. Use of Phosflow buffer from BD Biosciences allows better recovery of cells as compared to 90% methanol.

Acknowledgments

The author would like to thank GCRC for providing the samples of young control subjects. This study is supported by grant AG027512 and new scholar grant from the Ellison Medical Foundation.

References

1. Banchereau J, Steinman RM (1998) Dendritic cells and the control of immunity. Nature 392:245–252
2. Melief CJ (2008) Cancer immunotherapy by dendritic cells. Immunity 29:372–383
3. León B, Ardavín C (2008) Monocyte-derived dendritic cells in innate and adaptive immunity. Immunol Cell Biol 86:320–324
4. Ouaaz F, Arron J, Zheng Y, Choi Y, Beg AA (2002) Dendritic cell development and survival require distinct NFκB subunits. Immunity 16:257–270
5. Barnes PJ, Karin M (1997) Nuclear factor-kappaB: a pivotal transcription factor in chronic inflammatory diseases. N Engl J Med 336:1066–1071
6. Ardeshna KM, Pizzey AR, Devereux S, Khwaja A (2000) The PI3 kinase, p38 SAP kinase, and NFκB signal transduction pathways are involved in the survival and maturation

of lipopolysaccharide-stimulated human monocyte-derived dendritic cells. Blood 96: 1039–1046

7. Neumann M, Fries H-W, Scheicher C et al (2000) Differential expression of Rel/NFκB and octamer factors is a hallmark of the generation and maturation of dendritic cells. Blood 95:277–285
8. Agrawal A, Agrawal S, Cao J, Su H, Osann K, Gupta S (2007) Altered innate immune functioning of dendritic cells in aging humans: Role of PI3Kinase signaling pathway. J Immunol 178:6912–6922
9. Agrawal A, Tay J, Ton S, Agrawal S, Gupta S (2009) Increased reactivity of dendritic cells from aged subjects to self antigen, the human DNA. J Immunol 182:1138–1145
10. Romani N, Gruner S, Brang D, Kämpgen E, Lenz A, Trockenbacher B, Konwalinka G, Fritsch PO, Steinman RM, Schuler G (1994) Proliferating dendritic cell progenitors in human blood. J Exp Med 180: 83–93
11. Cella M, Sallusto F, Lanzavecchia A (1997) Origin, maturation and antigen presenting function of dendritic cells. Curr Opin Immunol 9:10–16
12. Agrawal S, Agrawal A, Doughty B, Gerwitz A, Blenis J, Van Dyke T, Pulendran B (2003) Cutting Edge: Different TLR agonists instruct dendritic cells to induce distinct T-helper responses, via differential modulation of ERK MAP kinase and c-fos. J Immunol 171:4984–4989

Chapter 13

Preparation of Collagen-Coated Gels that Maximize In Vitro Myogenesis of Stem Cells by Matching the Lateral Elasticity of In Vivo Muscle

Tathagata Chaudhuri, Florian Rehfeldt, H. Lee Sweeney, and Dennis E. Discher

Abstract

The physical nature of a cell's microenvironment – including the elasticity of the surrounding tissue – appears to exert a significant influence on cell morphology, cytoskeleton, and gene expression. We have previously shown that committed muscle cells will develop sarcomeric striations of skeletal muscle myosin II only when the cells are grown on a compliant gel that closely matches the passive compliance of skeletal muscle. We have more recently shown with the same types of elastic gels that mesenchymal stem cells (MSCs) maximally express myogenic genes, even in the absence of tailored soluble factors. Here, we provide detailed methods not only for how we make and nanomechanically characterize hydrogels of muscle-like elasticity, but also how we culture MSCs and characterize their myogenic induction by whole human genome transcript analysis.

Key words: Mesenchymal stem cells, Myogenesis, Elasticity, Microenvironment, MyoD

1. Introduction

Cell culture has a principal goal of translating in vivo cellular environments to in vitro platforms that enable better control and deeper study. Growth factors in serum media as well as oxygenation are rightly considered important, but physical factors should not be neglected. For example, cell cultures are maintained at 37°C because many cellular processes prove highly sensitive to even small deviations from body temperature – noting that fever is defined by only a 1–2°C increase. Additional physical features of the in vivo microenvironment are also likely to be important. In particular, most tissue cells are taken from physically soft

Irina M. Conboy et al. (eds.), *Protocols for Adult Stem Cells*, Methods in Molecular Biology, vol. 621,
DOI 10.1007/978-1-60761-063-2_13, © Springer Science+Business Media, LLC 2010

tissues and yet most culture studies are conducted on rigid glass coverslips or polystyrene culture dishes. Synthetic gels can be made that mimic not only the extracellular matrix ligand but also the elasticity of soft tissues and thereby enable tests of the hypothesis that tissue elasticity affects cell fates.

In most soft tissues – brain, muscle, skin, etc. – a combination of adherent cells and extracellular matrix establish the relatively elastic environment. At the macroscopic scale, elasticity is evident from a solid tissue's ability to recover its shape within seconds after acute deformations (e.g., poking). Regardless of geometry, the intrinsic resistance of a solid to a mechanical stress is measured by the solid's elastic modulus E, which is most simply obtained by applying a force and measuring the resulting deformation. Based on such measurements, comparisons of three diverse tissues which contain a number of different cell types and distinct extracellular matrices show that brain tissue with all of its neurons and other cell types ($E \sim 0.5$ kPa (1)) is distinctly softer than unstimulated striated muscle ($E \sim 10$–20 kPa transverse to the muscle axis (2)). Muscle appears in turn to be softer than the combination of epithelial cells, fibroblasts, and other cell types that make up skin ($E \sim 50$ kPa (3)). Translating tissue Es to the types of elastic gel systems useful for cell culture is the focus here. Numerous gel systems – particularly polyacrylamide gels – have tunable E that can be adjusted over several orders of magnitude from extremely soft to stiff, thus mimicking a wide range of tissues by controlling the extent of polymer cross-linking.

Given the appropriate adhesive ligand at sufficient density, cells from solid tissues adhere to a wide-range of elastic matrices (4, 5). Adhesion complexes generally couple into the cell's actin–myosin contractile apparatus, which assembles into *stress* fibers in cells cultured on rigid substrates (6). What is new and important is that this prototypical assembly process sets up a feedback of matricellular elasticity on cytoskeletal structure and adhesions (7). Potent mechanotransductive components couple some well-known signaling processes into anchorage-dependent viability and additional processes such as differentiation (2, 8) as is emphasized here with adult bone-marrow-derived mesenchymal stem cells (MSCs). These cells have been found to differentiate into various cell lineages – nascent neurons, myocytes, and osteoblasts – when cultured in identical media on substrates with tissue-mimetic stiffness (8). The goal of this chapter is to summarize the key methods in making and characterizing the cultures that maximize elasticity-promoted myogenic induction of MSCs. The methods include gel preparation, mechanical measurements of gel elasticity, culture details, immunofluorescence for protein markers, and isolation–amplification–labeling of RNA from these cells on gels for whole human genome microarray studies of cells on gels. The goal is to help guide the development of better cell culture methods.

2. Materials

2.1. Cell Culture

1. Dulbecco's Modified Eagle's Medium (DMEM), low glucose 1× (Gibco, Invitrogen, Grand Island, NY) (1 g/L D-glucose, L-glutamine and 110 mg/L Sodium Pyruvate).
2. Fetal bovine serum (FBS, HyClone, Ogden, UT).
3. Penicillin/Streptomycin (Cellgro, Mediatech Inc, Herndon, VA).
4. Solution of trypsin (0.05%) and ethylenediamine tetraacetic acid (EDTA) (1 mM) (Gibco, Invitrogen, Grand Island, NY).
5. Trizol Reagent (Invitrogen, Carlsbad, CA).
6. Dimethyl sulfoxide, DMSO (Hybri-Max) (Sigma, St. Louis, MO).
7. Triton-X solution, 0.5% (Sigma, St. Louis, MO).

2.2. Polyacrylamide Gel Preparation

1. 25 mm cir. Microscope coverslips (Fisherbrand, Fisher Scientific, Pittsburgh, PA).
2. 25 mm sq. Microscope coverslips (Fisherbrand, Fisher Scientific, Pittsburgh, PA).
3. 45×50 mm Microscope coverslips (Fisherbrand, Fisher Scientific, Pittsburgh, PA).
4. 3-Aminopropyltriethoxysilane (APTES, Aldrich, St. Louis, MO). Moisture-sensitive reagent, corrosive).
5. Ammonium persulfate (APS) (Bio-Rad, Hercules, CA). 10% solution should be stored in small aliquots at −20°C.
6. *N,N,N′,N′*-Tetramethylene-diamine (TEMED) (Sigma, St. Louis, MO). Highly flammable and corrosive.
7. Acrylamide, 40% stock solution (Sigma, St. Louis, MO). Sterile filtered. Should be stored at 4°C and protected from light (not longer than 3 months after opening; see Note 1).
8. *N,N′*-Methylene bis-acrylamide (Sigma, St. Louis, MO).
9. Sodium hydroxide (Sigma, St. Louis, MO).
10. 50 mM HEPES buffer, pH = 8.5 (99.5%, Sigma, St. Louis, MO). Sterile filtered (see Note 2).
11. Sulfosuccinimidyl-6-(4′-azido-2′-nitrophenylamino) hexanoate (Sulfo-SANPAH) (Pierce, Rockford, IL). Solutions should be stored at −20°C.
12. Glutaraldehyde (50% Grade I, Sigma, St. Louis, MO). Should be stored at −20°C.

13. Dichlorodimethylsiloxane (CMS) (Aldrich, 99.5%, St. Louis, MO). Air sensitive, highly inflammable, and irritant.
14. Acetic acid, glacial (Fisher Scientific, Fair Lawn, NJ).
15. Collagen I (from rat tail, BD Biosciences, Bedford, MA). Should be stored at 4°C.
16. Distilled water (dH_2O).
17. Phosphate buffered saline (PBS) solution (Gibco, Invitrogen).
18. Dulbecco's Phosphate Buffered Saline, PBS (1×, no magnesium chloride, no calcium chloride) (Gibco, Invitrogen, Grand Island, NY).
19. 75 cm^2 (larger flasks may be used for initial cell expansion) canted neck cell culture flasks (Corning, NY).
20. 6-well tissue culture treated plates (Corning, NY).
21. Bone marrow derived adult human mesenchymal stem cells (hMSCs) (Lonza, Walkersville, MD).

2.3. Indirect Immunofluorescence for the Myogenic Proteins

1. Paraformaldehyde solution (PFA), 10% (32% solution, EM grade, Electron Microscopy Sciences, Hatfield, PA) in 1× PBS. Prepare a 4% (v/v) solution by diluting 1 part with 7 parts of 1× PBS. Vortex the solution for several minutes to make sure that the solution is homogeneous.
2. 0.5% Triton X (v/v) solution in 1× PBS for permeabilizing the cell membrane.
3. 1× Phosphate-buffered saline for the different washes.
4. 3% bovine serum albumin (BSA) (Sigma, St. Louis, MO) in 1× PBS-Blocking solution.
5. 1% BSA in 1× PBS- Primary/secondary antibody dilution buffer.
6. Primary antibodies used in the respective ratios as and when necessary. For example, MyoD (Sigma, St. Louis, MO) (1:100 dilution); Pax7 (Developmental Studies Hybridoma Bank, DSHB, Iowa City, IA) (1:200 dilution).
7. Alexa 488 conjugated donkey anti-mouse secondary antibodies (Molecular Probes, Invitrogen, Eugene, OR).
8. Rhodamine phalloidin solution in methanol, 6.6 μM (Molecular Probes, Invitrogen, Eugene, OR) for actin staining.
9. DAPI (4,6-diamidino-2-phenylindole) solution (Sigma, St. Louis, MO) for nuclear staining.
10. Aqueous Mounting Medium-Biomeda Mounting Media (Gel/Mount) (Biomeda Corp., Foster City, CA).
11. Nail polish (N.Y.C., Uniondale, NY).

2.4. Microarray Analysis of the MSCs Cultured on the Polyacrylamide Gel Substrates

1. Trizol reagent (Invitrogen, Carlsbad, CA) for storing the cell pellets prior to RNA extraction followed by the microarray analysis.
2. QIAzol lysis reagent (Qiagen, Maryland) contains phenol, guanidine thiocyanate: toxic, corrosive. This solution is used for efficient lysis of the cells for total RNA extraction.
3. Buffer RPE (Qiagen, Maryland).
4. WT-Ovation Pico RNA Amplification System kit (NuGEN Technologies Inc., San Carlos, CA).
5. WT-Ovation Exon Module (NuGEN Technologies Inc., San Carlos, CA).
6. FL-Ovation cDNA Biotin Module V2 (NuGEN Technologies Inc., San Carlos, CA).
7. Affymetrix Human Gene 1.0ST Chip (Affymetrix, Santa Clara, CA).

3. Methods

In order to obtain consistent results with these polyacrylamide gels, it is critically important to maintain the polymerization conditions, buffer parameters, cross-linking, the number of cells (mesenchymal stem cells, MSCs) plated and the number of passage of cells used for plating on the gels of desired stiffness. It is very important to determine the correct concentration of the cross-linker as well as other components.

3.1. Cleaning and Functionalization of Glass Coverslips

1. Cleaned 25 mm coverslip(s) are placed on a hot plate (Corning). These coverslips can be sprayed with 70% alcohol and dried using Kimwipes just before use in order to get rid of all the accumulated dust. 0.033 N NaOH is prepared and the solution is filter-sterilized before use. Freshly prepared solution is recommended at this step. 900 μl of 0.033 N NaoH is carefully pipetted on to the coverslip. This ascertains a uniformity of the level of the solution on top of the coverslip, covering the entire glass surface. Then these coverslips are heated to 150°C till the liquid has completely evaporated from the glass surface. This evaporation should leave a thin white film on these coverslips.
2. These round coverslips are then removed from the hot plate and placed in the fume hood. Using a syringe, 120–150 μl of APTES is added slowly onto the surface of the coverslip(s) (see Note 3). It is to be noted here that not an excess of APTES is desirable at this step. APTES is added just to have a uniform covering of the glass surface. About 5–7 min is

allowed for APTES to react. This ensures a complete aminosilanization of these coverslips. At this point, coverslips are removed. Distilled water is sprayed at an angle to these coverslips, not too hard, allowing the water to trickle down along the surface. This step is to be continued until the point the water does not stick to the coverslips anymore. The aminosilanized coverslips are immersed in distilled water in 100 mm petri dishes. Generally, these coverslips are put in such a way so that they are not on top of each other. It is also to be noted here that great care should be taken to make sure that the treated side is up and it is not flipped around.

3. These coverslip(s) are rinsed with dH_2O thrice for 5 min each time.
4. They are then moved to a premixed solution of 0.5% glutaraldehyde in 1× PBS (stock concentration being 50% glutaraldehyde) in 100 mm petri dishes and allowed to stand for 30 min at room temperature. During this step, great care should be taken to make sure that the treated side of the coverslips is free and not covered by each other.
5. These coverslips are then rinsed again with dH_2O thrice for 5 min each time.

3.2. Preparation of the Polyacrylamide Gels

1. The edges of the aminosilanized round coverslips should be wiped with Kimwipes to dry and then they should be kept on a moist surface, preferably, wet paper towels (to avoid wrinkles of the PA gels at the edges due to evaporation), with treated side up. The top surface should be dry before the PA gel solutions are added.
2. Cold CMS solution is spread onto the 25 mm square coverslips (see Note 3). Care should be taken so that the volume added is just enough to cover the surface and addition should always be done in the fume hood as it is very corrosive. It is then allowed to stand for 1 min. The CMS is then wiped off from the coverslips with a kimwipe (see Note 4). dH_2O is then sprayed to these CMS-treated coverslips and wiped with kimwipes thrice. It is to be noted here that water should not be sticking to these square coverslips and excess CMS may lead to that. That is why removal of the excess CMS is very critical. These coverslips are then sprayed with dH_2O one more time and then set aside to dry, with the treated side up.
3. Acrylamide and bis-acrylamide solutions are made to their desired concentrations in 1× PBS to attain desired matrix elasticity. For 11 kPA gels, a solution of 6% acrylamide and 0.14% of bis-acrylamide in PBS should be prepared. However, it is essential to confirm the elasticity of the resulting hydrogels by AFM as explained later.

4. 1 ml of the desired solution is made. To 990 μl of this 11 kPA matrix elasticity solution, 10 μl of 10% APS (1/100 volume) and 1 μl of TEMED (1/1,000 volume) are quickly added and thoroughly vortexed for 5 s to obtain the PA gel solution.
5. 35 μl of this gel solution is subsequently (avoid delays since the polymerization starts immediately) pipetted onto the center of the aminosilanated coverslips (as described in the procedure for preparing coverslips) with the treated size up. Then, to a flat and uniform gel surface, CMS-treated square chlorosilanated coverslips, prepared in step 2 of this procedure are carefully placed on top of the polymerizing gel solution, with the treated side down. Great care should be taken here in order to avoid air bubble formation due to the sandwich formation between the two coverslips. Air bubbles will lead to a nonuniform cell culture surface and erroneous measurements of the matrix elasticity by AFM. The glass slips are covered with the top of a six-well plate to keep the gels in a humid atmosphere to minimize evaporation effects on the edges. The remaining portion of the PA solution may be kept in the vial next to the gels just to confirm when the polymerization process is complete.
6. Polymerization of the gels is to be allowed for about an hour and a half (depending on temperature, APS quality, etc.). It is to be noted here that the softer the matrices, the longer time it will take to undergo polymerization.
7. The top coverslip is then removed very carefully and the coverslips with the gels are immediately immersed in 2 ml of 1× PBS in a 6-well plate, each of them being immersed in each of the wells. They are rinsed thrice with 2 ml of 1× PBS to remove the excess unpolymerized acrylamide.

3.3. Procedure to Attach Collagen

1. 1× PBS is then removed from the 6-well plates. Using tweezers, these gel-coated coverslips are taken out of the 6-well plates and put in new, sterile 6-well plates, keeping the gel-coated side up.
2. Sulfo-SANPAH solution is prepared. 50 mg of the powder is dissolved in 50 ml 50 mM HEPES, pH = 8.5 buffer. This gives a concentration of 5×. This solution is then diluted 5× to 1× and then small volumes are aliquoted and frozen at −20°C. Frequent freeze-thawing of this solution is avoided and protected from light. Approximately 300 μl of the Sulfo-SANPAH solution (1×) is then added onto the gel surface preferably in the dark just enough to cover the gel surface on the coverslips. They are then exposed to 365 nm UV light (Spectroline, Westbury, NY) for 20 min until the color of Sulfo-SANPAH turns to rust brown. The phenylazide group of Sulfo-SANPAH, upon photoactivation at 365 nm, covalently attaches to the gel, leaving the succinimidyl ester free to react with amino groups of collagen or other proteins.

3. The PA gels are then rinsed with 50 mM HEPES, pH = 8.5 buffer thrice. These washes are preferably done in the dark.
4. Fresh collagen solution is prepared by diluting the stock collagen available from BD Biosciences in ice-cold 0.1 M acetic acid to a final concentration of 2 mg/ml. This solution is then diluted 10 times in ice-cold 50 mM HEPES, pH = 8.5 buffer, to yield a final concentration of 0.2 mg/ml.
5. 1 ml of the collagen solution is added to cover the gel surface on the coverslips in each of the wells of the 6-well plate and kept on a rocker at 37°C overnight in dark. Depending on the stock collagen concentration, overnight incubation conditions are sometimes altered to avoid collagen precipitation. Inucbations are also carried out at 4°C in dark when the stock collagen concentration is high (> 3.25mg/ml).
6. The next morning, the collagen-coated coverslips are rinsed with 1× PBS thrice, 5 min each time at room temperature in sterile conditions and subsequently transferred to a sterile 6-well plate.
7. 1.5 ml of 1× PBS is added to each of these wells so as to cover the gel-coated coverslips and put in tissue culture hood for at least 90 min under UV illumination by a germicidal lamp with the lid upside-down next to the plate. Care should be taken about PBS not evaporating and leaving the gels dry.

3.4. Growing and Maintenance of the Human Mesenchymal Stem Cells and Plating of These Cells on the PA Gels

1. Human MSCs are grown in DMEM media (low glucose) supplemented with 10% FBS and 1% Penn/Strep at 37°C and 5% CO_2.
2. The medium is changed every other day for conducting the cell on gel experiments.
3. At about 50–60% confluency, which is attained within 4–5 days of plating, the cells are ready to be split (see Note 5).
4. Cell culture media is aspirated out and the flask is washed with 1× PBS once.
5. These cells are then trypsinized using the Trypsin–EDTA and the flask is kept at 37°C incubator for 3–4 min.
6. The cell culture medium was added into the flasks and then the cells were pelleted by centrifugation at 5,000 rpm (1,500 g) for 5 min.
7. The medium is aspirated out and fresh medium is added and cell counting is done using the hemocytometer.
8. The cells are then split (1:4) for either replating in the flasks or on the gels (see Note 6).
9. For plating these MSCs on gels, ~2,000 cells are plated on each of the 25 mm circular coverslips coated with 11 kPa polyacrylamide (PA) gels (see Note 7).

10. The cells on gels are then cultured at 37°C and 5% CO_2 for different time points ranging from 3 days to 2 weeks, depending on the experimental conditions.

3.5. Measuring the Young's Elastic Modulus E by AFM

Amplifying the importance of good microscale measurements of elasticity are in vitro findings that even small microenvironmental variations – perhaps just 20–25% change in matrix stiffness – can have profound effects on the differentiation processes cited above as well as other cell functions. Since the Young's elastic modulus E is highly sensitive to polymer mass and crosslinking but is also the key parameter in mimicking the in vivo muscle environment for optimal myogenesis, it is essential to precisely determine the surface mechanical properties of the PA gels by atomic force microscopy (AFM). Nowadays, there are many commercial AFMs available that are suited to measure the elasticity E of soft samples by recording force-indentation curves. Many of the instruments even posses an $x-y$ scanner to raster the surface at high resolution but this is not mandatory to confirm the elasticity of a homogeneous sample such as the PA gels. However, for inhomogeneous samples such as native tissue or cells on gels, lateral elasticity maps can be obtained that can help to elucidate the micromechanics of these systems. Two types of probe geometries are especially suited to measure the micro-elasticity on the same length scale a cell experiences its environment, namely pyramidal tips (such as Veeco's DNP cantilevers and the corresponding Olympus OMCL-TR 400 PSA) and cantilevers with spheres attached (e.g., PT.BORO.SN.5 from Novascan). Pyramidal shaped probes are very commonly used and therefore relatively cheap. Spherical probes are more difficult to produce and thus more expensive. It is essential that the spring constant K of the cantilevers is soft enough (ranging from 5 to 100 pN/nm) to measure elasticities of the order of 1–100 kPa. As sketched in Fig. 1a, the cantilever with the probe at its tip is indented into the sample and the deflection measured by the reflected laser light on a photodiode. The applied force F is the product of deflection d and the spring constant K of the cantilever. This way force-indentation curves can be recorded to measure the elasticity of the sample.

For a detailed protocol of how to conduct force-indentation experiments in contact mode, the reader is referred to the respective instructions of the AFM manufacturer. In the following, we note some generally important issues.

When setting up the AFM measurements, great care should be taken with the calibration of deflection sensitivity. Since the spring constant K of the cantilever stated by the manufacturer is only a nominal value, each cantilever should be calibrated before measurement. From our experience, using the thermal method (9) in air is the best way to obtain a reliable value of the cantilever stiffness.

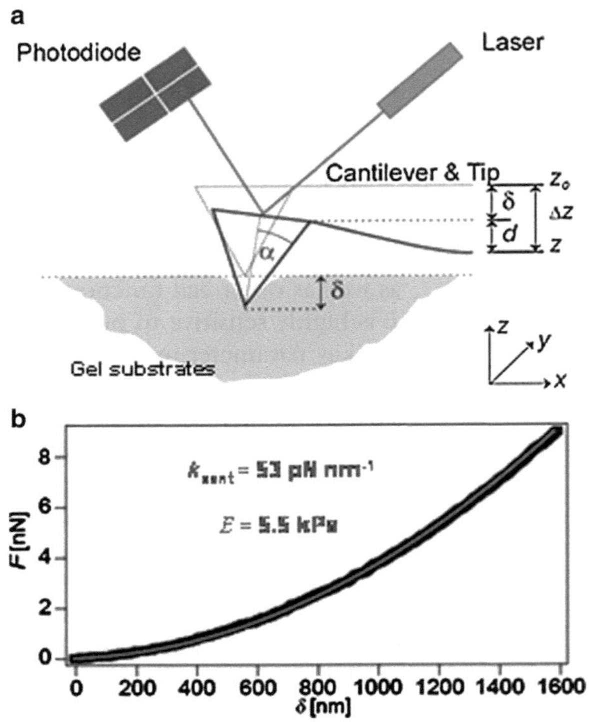

Fig. 1. Measurement of the elasticity *E* by AFM (**a**) The cantilever with a pyramidal tip (opening angle, α) is pressed down (Δ*z*) into the sample, leading to a certain indentation $\delta = z - d$ and causing the deflection *d* measured by the photodiode. (**b**) The Young's modulus *E* is determined by analyzing the resulting force indentation curves with a modified Hertz model. Black open circles are data points and solid red line is the best fit in the deflection range from 10 to 100 nm

After these initial steps, one should not forget to determine the deflection sensitivity in the liquid environment. The PA gels should be always measured and stored in PBS solutions to avoid problems due to changes in osmolarity. After mounting the sample, one should wait at least 10 min for thermal equilibration of the system. Then, force-indentation curves can be recorded.

Here, it is important to have a sufficiently long and flat baseline (at least 1–2 μm) while the cantilever is approaching the sample surface. The indentation velocity should not exceed 5 μm s^{-1} to avoid artifacts. After contact with the sample, the tip should be pressed into the sample to reach a deflection of at least 100 nm to obtain data that can be fitted well with a modified Hertz model (10–12). For best results, only the data in the deflection regime from 10 to 100 nm should be included in the fit as depicted in Fig. 1b.

The equation for the pyramidal geometry is:

$$E = \frac{\pi(1-\upsilon^2)F}{\delta^2 \, 2\tan\alpha},$$

where $F=kd$ is the applied force, $\delta = z-d$ the indentation of the probe, ν Poisson's ratio and α the opening angle of the pyramid (usually 35°).

For a spherical probe, the equation modifies to:

$$E = \frac{3(1-\upsilon^2)F}{4\delta^{3/2} R^{1/2}},$$

where R denotes the radius of the sphere.

In practice, neither the baseline deflection nor the contact point of the tip on the sample will be zero. Therefore, the following equation (for the pyramidal geometry) should be used to fit the Young's modulus E together with contact point z_0. Here, d_0 is the average deflection value of the baseline.

$$z = z_0 + (d-d_0)\sqrt{\frac{\pi k d(1-\upsilon^2)}{2\tan\alpha E}}$$

For sufficient statistics, 10 force-indentation curves per location should be taken and at least ten different locations on the hydrogel should be probed. To confirm the mechanical properties of the PA gels, one extra test sample for AFM should be prepared with each experiment.

3.6. Immunofluorescence Quantitation of Protein Expression

1. After the desired time points of culturing the cells on gels, the culture media is aspirated out and discarded.
2. The cells on the gels are then rinsed gently twice with 1× PBS.
3. The cells are then fixed in 10% PFA in 1× PBS for 15 min under gently rocking conditions.
4. PFA is then discarded and subsequent washes of these fixed cells are carried out. The cells on gels are now washed twice with 1× PBS for 5 min each time under gentle rocking conditions.
5. 0.5%-Triton-X 100 (solution in 1× PBS) is added to these cells to permeabilize these cells and then incubated for 15 min.
6. These cells are then rinsed twice with 1× PBS, 5 min each time.
7. 3% BSA (solution in 1× PBS) is added to each of the samples in the 6-well plates for blocking and incubated under gentle rocking conditions for 1 h.

8. The blocking solution is then removed and the primary antibody solution, mouse monoclonal anti-MyoD1 (1:100 solution in 1% BSA in 1× PBS) (Sigma) or mouse monoclonal Pax7 antibody (DSHB) (1:200 solution in 1% BSA in 1× PBS), is added. It is then incubated for 2 h at room temperature.
9. After removing the primary antibody, washing step is performed with 1× PBS thrice, 5 min each wash.
10. The secondary antibody solution is added next. Alexa Fluor 488 conjugated donkey anti-mouse secondary antibody (1:500 solution in 1% BSA in 1× PBS) (Molecular Probes) is used and the cells are incubated for 1 h.
11. Rhodamine–Phalloidin solution (1:400) is added at the same time during incubation with the secondary antibody.
12. 10 min before the incubation time is complete, DAPI is added to the wells for nuclear staining.
13. After 1 h of incubation, the samples are washed with 1× PBS thrice, 5 min each wash.
14. One wash is done with 0.5% Triton-X 100 solution in 1× PBS for 3 min.
15. Triton-X 100 is discarded and three more 1× PBS washes are carried out, 3 min each wash.
16. The last four washes are done with dH_2O, 4 min each wash.
17. The edges of the coverslips are then dried using Kimwipes before they are ready to be mounted on the 45×50 mm square coverslips.
18. A drop of the mounting medium, gelmount is added on the middle of the square coverslip.
19. Care should be taken when the round coverslips are inverted and kept on the square coverslip with the gelmount in the middle. Bubbles should be avoided during this step to avoid problems during fluorescence microscopy.
20. Nail polish is added just around the mounted and inverted round coverslip for sealing the samples of interest.
21. These slides are now either stored in dark at 4°C or viewed immediately.
22. They are viewed using fluorescence microscopy. Excitation at 488 nm induces the MyoD or Pax7 fluorescence (green emission), 554 nm excites the fluorescence of rhodamine–phalloidin (red emission) and excitation at 364 nm excites the fluorescence of DAPI (blue emission).
23. Quantitative imaging (Fig. 2) is done by imaging samples under conditions of constant excitation and camera gain and

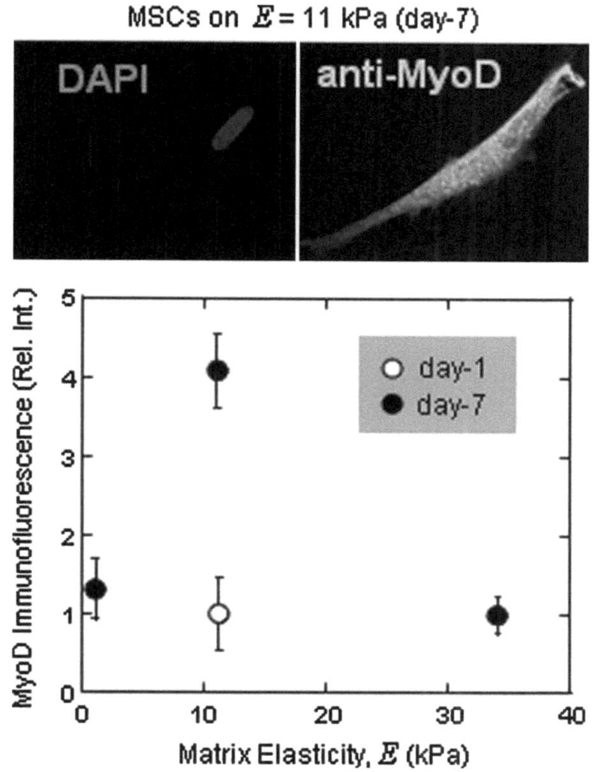

Fig. 2. Indirect immunofluorescence of myogenic markers in human mesenchymal stem cells after plating them on 11 kPA polyacryalmide gels for 1-day or 7-days. MSCs on the myogenic matrices (11 kPA gels) were nuclear stained with DAPI to fix the focal plane and probed with the primary mouse MyoD antibody and the corresponding Alexa 488 conjugated donkey anti-mouse secondary antibody. The plot shows the relative intensity of MyoD as analyzed per the elasticity *E*

all on the same day for at least 50 well-separated cells per sample. Focusing on the nucleus first standardizes the focal plane. Average intensity of cells is determined for cells immunostained with secondary antibody alone (I_0), and this is subtracted from the mean intensity (I) of images that are also immunostained for MyoD. This intensity difference can then be normalized to intensities of day-1 cells or cells on other gels. Day-1 cells are a preferred normalization because cells tend to have similar shapes at day-1 and day-7, and additional studies suggest little to no difference in MyoD between day-0 and day-1 cells.

3.7. Transcript Profiling

A. *RNA Isolation*

1. The cells from either the tissue culture flasks or the PA gels are harvested.

2. The cell culture medium is aspirated first and then the cells are washed with 1× PBS.
3. PBS is aspirated and 0.05% trypsin-EDTA is added.
4. The cells are incubated for 3–4 min in the 37°C incubator.
5. Cell culture medium is then added to the 6-well plates or the tissue culture flask and shaken gently to enhance the cell detachment.
6. The cell culture medium with the trypsin–EDTA is then transferred to Falcon tubes and centrifuged at 5,000 rpm for 5 min.
7. The supernatant is aspirated and fresh cell culture medium is added and the cell counting is done using hemocytometer.
8. Another centrifugation step is carried out at 5,000 rpm for 5 min.
9. Care should be taken this time so that the supernatant is completely aspirated (see Note 8). The cell pellets are stored in Trizol at −20°C prior to lysis.
10. The cells are lysed by adding QIAzol Lysis Reagent (Qiagen). For the pelleted cells, the cell pellets are loosened by flicking the tube and then pipetted to mix to ensure no visible cell clumping.
11. The cells are then homogenized by vortexing for 1 min (see Note 9).
12. The tube containing the homogenate is kept at room temperature for 5 min. This step promotes dissociation of nucleoprotein complexes.
13. Chloroform (140 μl) is added to the tube containing the homogenate and capped securely. This tube is shaken vigorously for 15 s. Thorough mixing is important for subsequent phase separation.
14. This tube containing the homogenate is stored at room temperature for 2–3 min.
15. Then, centrifugation is done for 15 min at 12,000 × *g* at 4°C. After centrifugation, the sample separates into three phases: an upper, colorless, aqueous phase containing RNA; a white interphase; and a lower, red, organic phase.
16. The upper aqueous phase is transferred to a new collection tube (supplied from Qiagen). 1.5 volumes (~525 μl) of 100% ethanol is added and mixed thoroughly by pipetting up and down several times. A precipitate may form after addition of ethanol, but this will not affect the procedure.
17. Immediately after this step, 700 μl of the sample is pipetted, including any precipitate that may have formed into an RNeasy

Mini spin column in a 2 ml collection tube (Qiagen). The lid is closed gently and centrifuged at 10,000 rpm for 15 s at room temperature. The flow-through is discarded at this step.

18. This step is repeated using the remainder of the sample and the flow-through is again discarded.
19. 500 μl of Buffer RPE (Qiagen) is added onto the RNeasy Mini spin column. The lid is closed gently and centrifuged for 15 s at 10,000 rpm to wash the column and the flow through is discarded.
20. 500 μl Buffer RPE is again added onto the RNeasy Mini spin column and the same step is repeated.
21. The RNeasy Mini spin column is transferred to a new 1.5 ml collection tube (Qiagen; and see Note 10). 50 μl RNase-free water is pipetted directly onto the RNeasy Mini spin column membrane. The lid is closed gently and centrifuged for 1 min at 10,000 rpm to elute the RNA (see Note 11).

B. *RNA Amplification*

This procedure is carried out according to the manufacturer's protocol using the WT-Ovation Pico RNA Amplification System (NuGEN Technologies Inc.) followed by the WT Exon Module (NuGEN Technologies Inc.). Briefly the steps are mentioned below.

1. First Strand cDNA Synthesis is done using the first strand reagents supplied in the kit. The first strand master mix is first prepared as described in the protocol.
2. Second Strand cDNA Synthesis is done using the second strand reagents supplied in the kit. The second strand master mix is then prepared.
3. Purification of the double-stranded DNA is carried out next using the RNA Clean beads supplied in the kit.
4. SPIA amplification is carried out with the supplied SPIA amplification reagents.
5. Amplified SPIA cDNA is purified at the last step using nuclease-free water provided with the kit.
6. Generation of ST-cDNA (sense-target) is performed by making a ST-cDNA Master mix solution as described in the protocol from the supplied Exon Module reagents.

C. *RNA Labelling and Fragmentation*

This procedure is also carried out according to the manufacturer's protocol using the FL-Ovation cDNA Biotin Module V2 (NuGEN Technologies Inc.).

1. Fragmentation of the SPIA cDNA is carried out with the supplied reagents first.

2. Biotin labeling of SPIA cDNA is done by first making the labeled master mixes as described in the protocol and then mixing the labeling master mixes with the fragmented cDNAs.
3. The fragmented and labeled cDNA is either stored at −20°C or prepared for gene chip array hybridization.

D. *Microarray*

The array is done with a Affymetrix Human Gene 1.0ST (Affymetrix) chip. It is known as the GeneChip Human Gene 1.0 ST Array. The hybridization, washing, staining and scanning of the chip is carried out following the protocols as described by the manufacturer.

4. Notes

1. The concentrations of the PA solutions should be consistent for the series of experiments. Longer storage (>3 months) of the PA stock solutions is not recommended due to degradation effects that can significantly influence the resulting gel elasticity, *E*.
2. All solutions used to prepare the PA gels should be sterile filtered. Freshly prepared solutions are always recommended. Otherwise, contamination is very likely to occur when cells are plated on these gels and incubated in the 37°C cell culture incubator.
3. APTES and CMS should always be applied to the coverslips in the fume hood as these chemicals are highly toxic and corrosive.
4. Extreme care should be taken while wiping the edges of the coverslips or during transferring from one plate to another with tweezers as touching of the gel surface may lead to contamination of cells on the gels when plated and incubated.
5. Care should however be taken that these cells do not touch each other in the cell culture flasks. About 50–60% confluency is ideal for these cell types to be trypsinized and replated or split. This is attained within 4–5 days of plating.
6. The number of passages given to these cell types is also a critical factor for studying the myogenic lineage of these cells when plated on PA gels. A maximum of 10 passages is generally recommended.
7. These cells are grown on 11 kPa gels for different time points of interest to determine the expression of specific myogenic markers like MyoD, Pax7, etc.

8. Incomplete removal of cell-culture medium during RNA extraction will inhibit lysis and dilute the lysate, affecting the conditions for binding of RNA to the RNeasy Mini spin column membrane. Both effects may reduce RNA yield. Incomplete loosening of the cell pellet during RNA isolation may lead to inefficient lysis and reduced RNA yields.
9. Incomplete homogenization during RNA isolation from the cell pellets may lead to significantly reduced RNA yields and can cause clogging of the RNeasy Mini spin column.
10. During the final step of addition of RPE buffer, following centrifugation, the RNeasy Mini spin column should be removed from the collection tube carefully so that the column does not contact the flow-through. Otherwise, carry-over of ethanol may occur.
11. If the expected RNA yield is >30 μg, the last step of the RNA isolation should be repeated with a second volume of 50 μl RNase-free water and eluted into the same collection tube. To obtain a higher total RNA concentration, this second elution step may be performed by using the first eluate. The yield will be 15–30% less than the yield obtained using a second volume of RNase-free water, but the final concentration will be higher.

Acknowledgments

Support of NIH NIAMS, NHLBI, NSF (DED), Association Francaise Contre Les Myopathies, France (TC), and the Humboldt Fdn. (FR) are gratefully acknowledged.

References

1. Gefen A, Margulies SS (2004) Are in vivo and in situ brain tissues mechanically similar? J Biomech 37:1339–1352
2. Engler AJ, Griffin MA, Sen S, Bonnemann CG, Sweeney HL, Discher DE (2004) Myotubes differentiate optimally on substrates with tissue-like stiffness: pathological implications for soft or stiff microenvironments. J Cell Biol 166:877–887
3. Diridollou S, Patat F, Gens F, Vaillant L, Black D, Lagarde JM, Gall Y, Berson M (2000) In vivo model of the mechanical properties of the human skin under suction. Skin Res Technol 6:214–221
4. Engler A, Bacakova L, Newman C, Hategan A, Griffin M, Discher D (2004) Substrate compliance versus ligand density in cell on gel responses. Biophys J 86:617–628
5. Rajagopalan P, Marganski WA, Brown XQ, Wong JY (2004) Direct comparison of the spread area, contractility, and migration of balb/c 3T3 fibroblasts adhered to fibronectin- and RGD-modified substrata. Biophys J 87:2818–2827
6. Buckley IK, Porter KR (1967) Cytoplasmic fibrils in living cultured cells. A light and electron microscope study. Protoplasma 64:349–380
7. Pelham RJ, Wang Y-L (1997) Cell locomotion and focal adhesions are regulated by substrate flexibility. Proc Natl Acad Sci USA 94:13661–13665
8. Engler AJ, Sen S, Sweeney HL, Discher DE (2006) Matrix elasticity directs stem cell lineage specification. Cell 126:677–689
9. Hutter JL, Bechhoefer J (1993) Calibration of atomic-force microscope tips. Rev Sci Instrum 64:1868–1873

10. Hertz H (1881) Ueber die Berührung fester elastischer Körper. J für die reine und angewandte Mathematik 92:156–171
11. Radmacher M (2002) In: Atomic force microscopy in cell biology, vol 68, pp 67–90
12. Dimitriadis EK, Horkay F, Maresca J, Kachar B, Chadwick RS (2002) Determination of elastic moduli of thin layers of soft material using the atomic force microscope. Biophys J 82:2798–2810

INDEX

F

G

H

I

L

Y

MIX
Papier aus verantwortungsvollen Quellen
Paper from responsible sources
FSC® C105338

If you have any concerns about our products,
you can contact us on
ProductSafety@springernature.com

In case Publisher is established outside the EU,
the EU authorized representative is:
Springer Nature Customer Service Center GmbH
Europaplatz 3, 69115 Heidelberg, Germany

Printed by Libri Plureos GmbH
in Hamburg, Germany